Advance Praise for
A Most Extraordinary Ride

"Marc Garneau has stood out as a leader since I first knew him as a young cadet. Despite facing demanding challenges, he projected a unique blend of optimism and fearlessness that served him well—both at sea and in outer space. Years later, we were together in the Liberal caucus, where he maintained that serenity and confidence, even in turbulent times. In *A Most Extraordinary Ride*, he tells us much more than how to become an astronaut, or a politician. With honesty and optimism, he shares how to rise, again and again—from life's falls and failures, pushes and punches—all the way to the stars."
—**Lieutenant-General (ret'd) The Honourable Roméo Dallaire**

"Marc Garneau is a national hero. Period. He has lived more life than most of us ever will, all of it deeply rooted in public service. Here he lays bare his two identities: the astronaut and the politician—one fixated on the challenges of space, the other, the challenges of making a better country. Garneau isn't afraid to confront the twists and turns of his own journey, ones we never knew about, while exposing the strengths and weaknesses of those around him. It is a blunt telling of a remarkable life both above earth and on it—a life fuller than even he may realize."
—**Peter Mansbridge**

"With characteristic humility and modesty, Marc Garneau leads us through his lifetime of adventure, accomplishment, and challenge in public service as a naval officer, astronaut, and politician. Along the way we learn the strength of resilience from his personal insights derived through setbacks, tragedy, and persistence. This is a masterwork of a principled public servant whose contributions demonstrate a role model for our time." —**Sean O'Keefe, NASA Administrator, 2001-05**

"This is the story of a courageous explorer, scientist, public servant, and patriot. I became a Marc Garneau fan and follower from the moment he became Canada's first astronaut. He is the rare example of someone who was excellent at every role he took on, from naval officer to astronaut to Parliamentarian. Humble, open, thoughtful and ever gracious, Marc's story should inspire pride in all Canadians. He is one of our best."

—Hon. John Manley

"Marc Garneau's 'extraordinary ride' gives us a front seat on his journey of purpose and public service. His story is a reminder that even in this age of cynicism, politics is still where good people can make a difference. Marc's courageous decision to harness his lifetime of accomplishment and bring to bear his intelligence, judgment, decency, integrity, and kindness into politics was a gift to all Canadians. I hope this story inspires other accomplished citizens to step up, and have the courage to put their names on ballots."

—Hon. Scott Brison, P.C.

MARC GARNEAU

A MOST EXTRAORDINARY RIDE

SPACE, POLITICS, AND THE PURSUIT OF A CANADIAN DREAM

SIGNAL
McCLELLAND
& STEWART

Library and Archives Canada Cataloguing in Publication
Title: A most extraordinary ride : space, politics, and the pursuit of a Canadian dream / Marc Garneau.
Names: Garneau, Marc, author.
Identifiers: Canadiana (print) 2024030084X | Canadiana (ebook) 20240300882 | ISBN 9780771016219 (hardcover) | ISBN 9780771016264 (EPUB)
Subjects: LCSH: Garneau, Marc. | LCSH: Astronauts—Canada—Biography. | LCSH: Politicians—Canada—Biography. | LCGFT: Autobiographies.
Classification: LCC TL789.85.G37 A3 2024 | DDC 629.450092—dc23

Jacket design by Andrew Roberts
Jacket art: National Archives and Records Administration
Typeset in Kepler by Terra Page
Printed in Canada

Published by Signal,
an imprint of McClelland & Stewart,
a division of Penguin Random House Canada Limited,
a Penguin Random House Company
www.penguinrandomhouse.ca

1 2 3 4 5   28 27 26 25 24

Penguin
Random House
Canada

To my children, as I promised.

A MOST EXTRAORDINARY RIDE

PROLOGUE

IT WAS JULY 20, 1969, and I was sitting in the cockpit of a sailboat somewhere off the coast of England, gazing intently at the Moon.

The boat, called the *Pickle*, of all things, and its crew of thirteen had just raced across the Atlantic Ocean from Newport, Rhode Island, to Cork, Ireland, and was on its way to London. Our navigator had the radio on, and we were listening, spellbound, as Neil Armstrong announced to the world, "The *Eagle* has landed."

Less than seven hours later, Armstrong and Buzz Aldrin would open the hatch of their lunar module, descend a small ladder, and plant their feet firmly on the Moon's surface. For the first time ever, humans were walking on another celestial body.

I kept looking up as I listened and could not help but think that here I was, in one of the oldest forms of human transportation, a sailboat, while astronauts had travelled through the vacuum of space and landed in the Moon's Sea of Tranquility. Like everyone else listening and watching that night, I was mesmerized, marvelling at the human ingenuity that had made something like that possible.

I understood the principles of orbital mechanics involved in spaceflight because at the time I was studying engineering physics at Kingston's Royal Military College. But the success of the Apollo 11 mission blew me away.

I had joined the Royal Canadian Navy at sixteen because I loved being out on the wide expanse of the ocean, a sometimes challenging place to be—not unlike space—where you had to rely on your wits and fend for yourself; a place where teamwork mattered. More than anything, it was that sense of freedom and self-reliance that had drawn me to my chosen career.

I had experienced it in the *Pickle* during the twenty days it took us to cross almost five thousand kilometres of ocean, sometimes in rough seas and bad weather, trimming and changing our sails to maximize our speed and navigating by the stars like the mariners of old. It had been nothing short of exhilarating and life-changing, much like the journey that has brought me to this point in time.

But that night, as I looked heavenward, I could not help but think (perhaps *dream* is a better word) that going into space would be extraordinary, perhaps the ultimate experience for any human.

I could not have known what the future held for me on that brilliant evening as *Pickle* sailed into the English Channel. Had I been told, I wouldn't have believed it.

I didn't set out to be an astronaut, but that's what I became; I also didn't set out to be a politician, but that also happened. What I did set out to do, and what this book is about, was to live to the fullest of my capabilities rather than shrink from the challenges life threw at me, to stay curious, and to carry myself with dignity. I'll let the reader be the judge of whether I succeeded.

ONE

I WAS A PRODUCT OF THE TWO CANADIAN "SOLITUDES." My father, André Garneau, was a French Canadian infantry officer from Quebec City and my mother, Jean Richardson, was an English Canadian nurse from Sussex, New Brunswick. They met in Montreal during the war, when my father was hospitalized before going overseas and my mother attended him. The relationship grew from there, even when he left to fight, and she was at Montreal's Windsor Station in 1945 when his train pulled in, bringing troops home from the war. They married the next year.

My father's side of the family had deep roots in Quebec, going back to 1656. My mother's English and Scottish roots, dating to the late eighteenth century, were in Nova Scotia and New Brunswick. I was half Quebecer and half Maritimer, a good combination that has served me well. I believe I get my passion and my tendency to argue from my Quebec ancestry and my pragmatism and can-do attitude from the Maritimer in my DNA.

My father was a principled man, guided by values that he passed on to me and my brothers. He held the traditional view (outdated today) that his role as our father was to be the primary custodian of his boys' upbringing. He was a straight arrow and always fair, although something of a disciplinarian, a product of his time and of his own upbringing, not to mention being a military man through and through. Although he didn't radiate empathy, he was generally good-natured and always available when we sought his advice. My mother, on the other hand, was ever approachable and the one we went to for sympathy and comfort. She was also extremely smart, as I would discover later in life, but tended to hide her light under a bushel. My father taught me how to be responsible. My mother taught me to dream of possibilities.

I was born in Quebec City at the Jeffery Hale Hospital, where my mother had trained as a nurse. I was the second child; my brother Braun got there thirteen months ahead of me. When I was eight years old, I would return to the Jeffery Hale for two months, with my left leg in traction after a skiing accident. One night, as I watched the local news on television, I saw, to my utter horror and disbelief, footage of smoke billowing out of my family's third-floor apartment on Brown Avenue. Worried I had lost everyone and would become an orphan, I screamed to the nurses to find out what had happened to my parents and brothers. Mercifully, they soon returned to tell me that everyone got out safely and they were staying with neighbours. Thankfully, Braun had smelt the smoke and woken up the rest of the family in time to get out. Quick, life-saving thinking on my brother's part. And thank goodness—I was certainly not prepared to become an orphan!

As a child, I had an innate curiosity, wanting to experience life head-on. Mix that with a preternatural self-confidence, sometimes to the point of stupidity, and I more than occasionally got into trouble. Once, Braun and I decided to mix two household products together because we had been told they would produce a sharp-smelling gas that would make people cough uncontrollably. Which is precisely what happened to me, at which point Braun woke my sleeping parents (he was good at that), shouting that I was dying! I wasn't, of course, but the event did not impress my parents one bit, and perhaps tempered some of my belief that I was invincible.

Ours was a Catholic home, and the family attended church every Sunday; I would become an altar boy and serve mass. I believed, in a literal way, everything the church taught me, to the point of developing a strong fear of going to hell if I committed a sin, something that gave me nightmares and that would shape my adult views of religion. Simply put, I don't believe religion should be based on fear.

Something else happened to me that would play a role in my future perception of the Catholic Church. For three summers, I attended a boy's camp, Camp Kinkora in the Laurentians, run by the Catholic Church, a wonderful experience except for one incident. One day, after I had been swimming, the camp director took me to his office and made me take off my swimsuit. I was ten or eleven at the time. As I recall, he only stared at me, all the while lecturing me for not being at dinner with the other boys. I could not understand why this happened and I kept silent about it, not even telling my parents, burying it deep in my memory. Many years later, this same director was charged with a sexual offence involving a minor, stemming from a different incident, and when it became public, my memory resurfaced. While I was able to put the incident behind me, you will understand my feelings about the Catholic Church.

Because Braun and I spoke English at home, our parents decided to send us to French school, so that we could be completely at ease in both languages. It was one of the great decisions they made in raising us, and I would go on to attend French school right through to university. Although I didn't know it at the time, being bilingual would open many doors for me.

When I was ten, I was shot by a friend. Considering the possibly huge consequences this could have had, I'm surprised I don't remember his name. His father had bought him a BB rifle for his birthday, and he could hardly wait to show it to me. We went to his house right after school, and I waited at the front door while he fetched it. A moment later he reappeared, pointed the rifle at me, and pulled the trigger. I felt a sharp sting below my left eye, which immediately blurred up. I panicked. My friend panicked. His father appeared out of nowhere and checked to see if I was all right (which I wasn't). Meanwhile his son was screaming that he didn't know the rifle was loaded and his father was shouting back, "I told you never to point that at anyone!"

I was rushed to hospital, where they froze the skin beneath my eye and removed the pellet. Fortunately, my vision gradually returned to normal. Less than a half an inch is all it would have taken for me to lose my eye and effectively set my life on a completely different trajectory. I don't know how many lives I have, but I used up one of them that day.

After the war, my father had decided to continue serving in the regular forces, a career from which he would retire as a brigadier-general. Because he was a serviceman, we moved a lot. By the age of twelve, I had lived in seven different places—from Quebec City to Oakville, back to Quebec, then to Germany and once more back to Quebec, where we lived in two different places, followed by Saint-Jean-sur-Richelieu for four years before moving to London, England. I was totally comfortable with pulling up stakes on a regular basis—probably a sign of my desire to experience the new and the unknown, and an inherent and lifelong restlessness.

Although I didn't know it at the time, Braun always being there made those moves much easier for me. Whenever we moved, I knew I would already have a friend at the other end. Braun was a constant and reassuring presence in my early life, my best friend. As young children, we shared the same room, sleeping in a bunk bed, which we often turned into a fort, hanging sheets around the lower bunk and pretending to fight some imaginary foe. We had the same friends and did absolutely everything together, playing football and hockey, going off on biking adventures, fishing off the pier on our holidays, or going to summer camp. We even got in trouble together, such as the (one and only) time that we stole a twenty-nine-cent bag of licorice from the grocery store and our mother caught us eating it and told my father, who marched us back to the store to confess our crime to the manager.

In 1961, when we learned of our move to London, where my father would work for NATO, there was great excitement in our household, which now included Charles, born in 1954, and Philippe, born in 1959. We would be crossing the Atlantic in a passenger liner, the *Empress of Britain*, my second voyage in a ship, having crossed the other way in 1955 after living in Soest, Germany, for two years when my father was deployed there as part of the Canadian postwar military presence in Westphalia.

That first voyage planted the seed that made me decide, nine years later, to join the navy. I was seven years old and found the experience mesmerizing. We crossed in the RMS *Samaria*, operated by Cunard. It had been built in 1920, and our crossing would be its last voyage before it was decommissioned. Its cruise speed was a paltry sixteen knots, which meant a longer trip than would be the case today, and the weather was rough for most of it. I would discover seasickness but also find my sea legs. I loved nothing more than to stand outside on one of the decks and watch the waves coming my way, with the wind blowing in my face as the ship pitched up and down, sometimes gently, sometimes violently. It was exhilarating. Eventually we entered the Gulf of St. Lawrence and made our way to Montreal, the destination of my first great sea voyage.

The three years I spent in London would be formative. I attended the French Lycée and made friends with students from many countries. This expanded my horizons in a way that few other experiences can. It also made me realize there was a big exciting world out there that I could hardly wait to explore.

In the summer of 1964, Braun and I set off on a month-long trip, making our way through France, Switzerland, and Italy in a quest to see the treasures of the Renaissance: Leonardo da Vinci's *Last Supper*, Michelangelo's statue of David, the Sistine Chapel, and so much more. Braun was sixteen and I was fifteen. It was more than our young minds, and eyes, could take in. But for me these encounters with such beauty sent my imagination reeling and only fuelled my curiosity.

In a courageous act of faith, our parents had purchased train tickets for us for the whole journey and had given us each three dollars per day (the equivalent of just under thirty dollars these days) to pay for our accommodation and meals and to buy a few souvenirs.

We stayed in youth hostels. Our journey took us from London to Basel, Zurich, Lucerne, Lugano, Como, Milan, Venice, Florence, Rome, Pisa, La Spezia, Genoa, Paris (where we ran out of money and slept in Saint-Lazare train station), and finally back to London. It was my first great adventure. We had done everything we had set out to do and made it back safely. That experience gave me the confidence to make it on my own. Unfortunately, the desire for independence it instilled in me, without the necessary maturity to go with it, would also get me into trouble.

Returning to Canada, our family moved to Camp Valcartier, forty-five minutes north of Quebec City and the home base of the Royal 22nd Regiment, where my father took command of the 3rd Battalion. The contrast could not have been greater. After living in one of the great cosmopolitan cities of the world with so much on offer, living on an army base felt like a punishment. I soon became restless and a little rebellious, longing to spread my wings and assert my independence.

On one occasion, against my parents' wishes, I went rabbit hunting in the middle of the wilderness with a friend, and our canoe tipped over in the Jacques-Cartier River. Night was falling on this cold November day, with ice already forming on the river's edge. After righting our canoe, we spent the next two hours paddling and then walking back home through the bush in the darkness, as hypothermia set in. It was a painful and scary experience, and I kept it from my parents for several years.

I attended school in Quebec City and was sometimes allowed to stay overnight at a friend's house. I was on the cusp of turning sixteen and most of my friends were a year or two older. Although we were all underage, we liked going out for a beer at a popular tavern in old Quebec called Le Colonial, where the waiters didn't check our age. Being an old-style tavern, the beer was free-flowing. Unfortunately, one time a friend and I drank way too much of it, teenagers not exactly being known for moderation. When we left the tavern, I was not in control of my faculties. Walking down the street, my friend spotted a parked car with an open trunk in front of the local radio station. Someone was unloading equipment and had gone into the station. My friend looked at me and said, "Watch this." He reached into the trunk and pulled out a piece of equipment. At that moment, the car's owner reappeared and started shouting at us. My friend ran off and I, not knowing what to do and quite drunk, took off in a different direction.

Within minutes I heard a police siren and quickly hid in the shadows on a little side street, panic-stricken and wondering whether I had been identified. I was scared, intoxicated, and fearing the worst. I waited ten minutes, and when things appeared quiet, went looking for my friend.

As I rounded a corner, two policemen spotted me and I froze, no doubt looking quite guilty. I submitted without resistance as they apprehended me and drove me to the police station, where I was asked some questions, including my parents' address and phone number. I was then taken to jail and locked in a cell, which already had another occupant.

Thus began one of the longest and most difficult nights of my life. I had been arrested on suspicion of theft, and although I did not actually commit the crime, I was with someone who did and I had subsequently fled the scene. This was no Hollywood movie. This was the real world, and being arrested, interrogated, and incarcerated shook me to the core. I feared I would be convicted and have a criminal record for the rest of my life. I had been monumentally stupid and would now suffer the consequences. What would happen to me? What would my parents think?

Fortunately, my cellmate, an adult, was not in the mood to talk. As I gradually sobered up and imagined all the worst possible outcomes, a policeman came by and told me that my father was on his way from Valcartier. At this point, I considered how my actions would embarrass him, the commanding officer of the 3rd Battalion of the Royal 22nd Regiment. In Quebec City, this was considered a prestigious position. Quebecers took immense pride in the R22R. I felt genuine shame.

Shortly after two in the morning, I was taken from my cell and escorted to the front of the station, where my father was waiting. The policeman told us we were free to leave. We got in the car and headed home. Neither of us spoke for about twenty minutes, until my father said: "What happened?" By this time I was sober and I did my best to recount the events that had led to my arrest. As I spoke, my father remained calm, listening as he drove. As we pulled into the driveway, he told me to get some sleep and that he and my mother would talk to me in the morning. He knew I was totally exhausted.

The next day, we had that talk and I answered all my parents' questions. I was expecting a stern lecture and some form of punishment, but both of them remained calm throughout, never raising their voices, perhaps realizing that I fully understood the stupidity and recklessness of my actions and that no more needed to be said. In hindsight, I realized they knew me well and I was grateful for their understanding. Most importantly, I acknowledged my guilt. I had been totally irresponsible. As I look back at what happened that night, I realize it could have been much worse had the police pulled their guns and I made some stupid move and bolted or who knows what.

Two months later, I appeared before a judge. My father and he spoke privately before I was ushered into his chambers. He was clearly used to speaking to young people who had erred, but he also wanted to make the point that, coming from an advantaged family, I had less excuse than others who had appeared before him. I was given a stern warning and told that I was fortunate that I would not have a criminal record, given my age. The judge also told me he would be severe with me if I ever appeared in front of him again. I left his chambers, deeply chastened. I had been given a second chance and I knew it.

To this day, I do not know what happened to my friend. By the time his case was heard that summer, I was no longer seeing him, and our paths have not crossed since.

My arrest and incarceration led to a period of deep introspection during which I made an important decision about my life. I was now sixteen years old and facing the fact head-on that I was undisciplined and immature. And yet I had so much going for me: my health, the capacity to learn, a loving family, and a bright future—provided I got my act together. I needed to anchor my life by finding its purpose, something that I could be passionate about and that would get me back on track. That spring, three factors drove me to make a decision: the desire to straighten myself out, the drive to set out on my own, and my father's own choice of career. I went down to the Canadian Armed Forces Recruiting Centre and joined the navy.

I had been preparing for this day my whole life, given my long-standing fascination with the ocean. The trigger, however, was my desire to regain my parents' trust after my run-in with the law. While this may sound a little like one of those recruitment ads from the Armed Forces, a career in the navy also offered me excitement and adventure and fulfilled my desire to serve my country. My father's father had fought in the Great War and been wounded at Lens and Passchendaele. My father, of course, had fought in World War Two, and although he never consciously pushed me towards a military career, his example and how he conducted himself gave me the desire to serve.

My first interview at the recruiting centre went well and I was told to return the following week for a physical exam and a review of my medical history. Everything was going well until I was asked whether I suffered from asthma, to which I replied, "Yes, I think so, when I was younger." The physician looked almost pained as he informed me that, unfortunately, a history of asthma disqualified me from joining the Armed Forces. I was crestfallen. My career in the navy was over before it started.

I went home and told my parents. Fortunately, they weren't ready to give up so fast. While they acknowledged using the word *asthma* in my childhood, they both thought that my breathing problems were the result of something else. They dug out my medical records and found that I had suffered episodes of acute bronchitis, but not asthma.

I returned to the recruiting centre, corrected the misunderstanding, and received the green light to be a candidate for military college, assuming I achieved satisfactory school marks. Had my parents not come to the rescue and cleared up the semantic misunderstanding, this could easily have been another one of those moments when my life took a different turn.

Later that summer, I joined 179 recruits at the Collège Militaire Royal de Saint-Jean, or CMR for short, choosing engineering as my field of study. Although this was a good choice for someone joining the navy, something else drove my decision: curiosity. I wanted to understand how things worked. When I turned the ignition key in a car, I wanted to understand how the car moved, from the starter to the engine to the transmission. I wanted to know how music came out of a radio, how wind was produced, how cows made milk from grass, how apples grew on trees. At school I had taken subjects like history, geography, literature, and the arts, and although I had quite enjoyed them, I felt I needed to pursue more technical subjects if I was to understand many more things, such as how airplanes fly, how electricity is produced, and how computers work.

My first year at CMR was extremely challenging, the equivalent of boot camp, and I was perpetually exhausted for months on end. A typical day began at 5:45, when my alarm went off so I could get dressed in time to run circles around the parade square, a form of punishment meted out liberally by senior cadets. This is when I took up smoking, because I was afraid I'd fall back to sleep unless I took a puff to jolt me awake. In fact, I would lay out the cigarette, lighter, and ashtray on my night table beside my alarm clock before I went to bed. (In those days you could smoke in your room.)

My first-year Christmas exam results put the fear of God in me and sent me into panic mode. I had ranked 162nd out of 180 students, in part because I kept falling asleep in class from exhaustion. (You could *not* smoke in class!) Unless I did something, I wasn't going to make it. I reached out for help and thankfully it was given. My worst subject was chemistry, and my teacher, Dr. Lavigne, offered to give me some one-on-one tutorials. Within weeks, it all began to make sense. As for my other subjects, I developed the discipline and the resolve to take them seriously, gradually turning things around. By year's end, I felt that I had caught up.

Attending university is demanding no matter where, but no more so than at a military college, with compulsory sports, military drills, and senior cadets shouting at you all the time. If you can hack it, it is without a doubt character-building, teaching you to be on time, to organize your life, and to live up to your obligations. It played a major role in shaping the person I became, although that's not to say that I wouldn't falter later on, as you will see.

I would remain at CMR for three years and then transfer to the Royal Military College, or RMC, in Kingston, Ontario, for my last two years before graduating. During the summer, I would train at Canadian naval bases in Esquimalt and Halifax.

It was during my second summer that I came within a whisker of being kicked out of the navy, and unfortunately it would not be the only time. One evening, while my ship, HMCS *Beacon Hill*, was at anchor, I and a couple of other cadets asked if we could go ashore in a Zodiac, a small rubber boat with an outboard motor. We wanted to visit the local sights, a not unreasonable request. The answer was an emphatic no.

After fuming for a while, the three of us hatched a plot: we would swim ashore! At the time, it did not occur to us to figure out the logistics of getting back to the ship afterwards. We didn't think that far ahead. Here is what we plotted. We would wait until dark, get into our bathing suits, and each fill a plastic bag with clothes for when we reached dry land. We would then sneak forward onto the fo'c'sle (the pointy forward end), clamber down the anchor chain, and swim ashore, plastic bags in hand. The shoreline was roughly three or four hundred yards from the ship.

Our plan was executed to near perfection, except for a couple of details. First, we had not anticipated that the anchor chain would be coated with slimy green algae, making it extremely slippery and therefore impossible to hold on to, causing us to let go and fall some distance into the water. Fortunately, no one got hurt and no one on the ship seemed to hear our splashes.

Once all three of us were in the water, we began to swim ashore in the dark, trying not to attract the attention of anyone who might be on deck enjoying the night air. All went well until we were about half-way to shore, when two large searchlights on our ship were turned on and began sweeping across the water in our general vicinity, gradually converging on our position. In desperation, we hid behind our plastic bags, hoping not to be spotted. Visions of the movie *The Great Escape* danced in my head.

When the searchlights fixed on us like two giant eyes, we knew the game was up. At this point, we saw no alternative but to continue our swim ashore, knowing that a Zodiac would be dispatched to pick us up. In some ways this was fortunate, since one of my friends was now shivering. Sure enough, the Zodiac arrived at a small dock at about the same time that we got out of the water. Looking and feeling like escaped prisoners, we boarded the craft, and minutes later we were back at the ship, where virtually the entire crew had gathered to view the returning escapees. It was quite the moment. While many of the sailors had smiles on their faces, the officers looked stern as we were escorted into an office and told that we had disobeyed a direct order and that there would be serious consequences.

So how did we get caught? When a ship anchors, it must ensure that the anchor does not drag along the bottom with the tides or strong currents, which could cause it to gradually move closer to shore and run aground. Consequently, the ship's position is monitored continuously on radar. We hadn't thought of that. When we began our swim ashore, the radar returns from our three floating bags, as well as our own heads and arms, appeared as a blip on the radar screen. When the blip slowly moved towards the shore, suspicion was aroused, and the decision was made to turn on the searchlights. A roll call of cadets was initiated, and the three missing culprits were quickly identified. As far as escape plots go, it was one of the worst ever concocted or executed. Let me blame it on my not yet fully developed frontal cortex.

Let me also say that my actions and those of my two colleagues were stupid and reckless. My punishment began with a severe warning. It was made clear to me that I had come very close to being kicked out of the navy and sent home. I was confined to my ship and given extra chores such as painting and cleaning. While my two friends would eventually leave the navy, I stuck with it. Despite my faltering start, I loved what I was doing.

The following summer, I trained in a minesweeper, and our ship made a weekend stopover in Boston. My aunt Norma and uncle Bert lived in Providence, Rhode Island, and I thought this might be a good time to visit. I placed a call to them, and they offered to pick me up in Providence if I could make my way there—which I did by hitchhiking, arriving about one in the morning on Saturday.

I had assumed I was free for the entire weekend, but in fact I was expected back at my ship that same night. While I was enjoying the next day with my relatives, an alert had gone out to all the police stations and hospitals in the Boston area, trying to locate me. The longer I was absent, the greater the concern was that something serious had happened to me.

After a wonderful visit, my relatives drove me back to Boston and I boarded my ship, completely unaware that I had provoked an international incident. I was told to report to the second-in-command, who looked at me as though I was an apparition and asked me whether

I realized the trouble I was in. It was only then that the penny dropped.

When I had climbed, or rather slid, down the anchor chain the previous summer, I knew that what I was doing was wrong and I did it anyway because I was annoyed (a childish reaction). In Boston, I was simply oblivious to what was expected of me. There was no malicious intent rooted in not getting my way or thinking that the rules didn't apply to me; rather, this was simply a case of not paying attention (which is also immature). As happened after the first time I "jumped ship," the navy decided not to throw me out, although later I learned that my commanding officer had strongly recommended they do so.

I look back at these slip-ups, both of which almost got me tossed out of the navy, and I honestly can't believe I behaved that way. It was like I had a switch in my brain that turned off every so often and made me abandon reason and responsibility. Clearly, I was learning some of life's most valuable lessons the hard way and was still a work in progress.

In the spring of 1969, I and seven of my RMC classmates received an extraordinary offer from Lt.-Cmdr. Jake Friel, our naval staff officer. We were asked if we would like to sail across the Atlantic Ocean in a fifty-nine-foot sailboat called the *Pickle*, which belonged to the Canadian navy and was sometimes used for training midshipmen. In this instance, we would be part of a crew of thirteen in a race from Newport to Cork, a voyage of almost five thousand kilometres. My answer was an immediate and resounding yes to this once-in-a-lifetime opportunity. To compete in a race across the ocean under sail would be an extraordinary adventure.

The *Pickle*, a sturdy wooden vessel built in 1936 for racing in the Baltic Sea, would be up against modern racing yachts made of aluminum and fibreglass and crewed by professionals. Our own crew would be composed of two watches of five people each, plus a cook, a navigator, and a skipper. As a watch, our primary responsibility was to change the sails, depending on sea state and wind conditions, both of which could change in strength and direction at any time, day or night. We also trimmed the sails continuously to maximize our speed. Because one watch was always on duty while the other rested or slept, we used the "hot bunking" system, which meant that the five of us used the same bunks as the other watch.

When we were on watch, my watchmates and I had to be ready to change a sail at any moment. Speed of execution was the goal, since the vessel would inevitably lose speed during sail changes. Three of us would go forward and retrieve the new sail from the sail locker, remove it from its bag, and position it to hoist it up as soon as the old sail was lowered. If conditions were rough, we tethered ourselves to the guardrails as we moved forward on a slanted deck. Our vessel could sometimes pitch forward suddenly, and occasionally violently, and anyone could, in the worst case, end up overboard, a particularly dangerous outcome if it happened at night and in rough seas when visibility was severely reduced. More than once, I proceeded forward cautiously on a slippery deck in the middle of the night with two crewmates to change a sail as the bow of the vessel dropped suddenly and a wall of water washed over us. The elements were merciless, and we sometimes returned to the cockpit completely drenched. I came to appreciate the all-too-rare pleasure of wearing dry socks.

Our crossing exposed us to a range of sea and weather conditions. From strong winds to light breezes, from small undulating swells to large waves, we experienced it all. We sailed through the edge of the Grand Banks of Newfoundland where the fog enveloped us, and because we had no radar, we had to post someone at the bow to listen and look for any other vessel we might encounter. This was an eerie experience, since your imagination can run wild when you're alone at the front of the boat, shrouded in fog, in the middle of the night.

Each one of the crew was expected to perform a specific task. This meant working as a team, knowing that your crewmates were depending on you, just as you were depending on them. I was always conscious that we were on our own and that if something went wrong, there was no guarantee of being rescued. We were left to our own devices, and this became obvious when you scanned the horizon and saw nothing but empty sea and sky.

At times, sailing across the ocean was mentally demanding and could affect your disposition. Everyone handled it differently. Most obvious was the fact that we were constantly living in close quarters and could never go off and be alone for a while. We were always together, whether working, eating, or sleeping. Simply put, there was no privacy at all. Although humans are social animals, we occasionally appreciate solitude. And given that we had no choice but to live in close quarters, sometimes in a difficult environment, some of my crewmates coped by becoming chattier while others became less so. In both cases, they were adapting as best they could.

How well we knew each other was also a factor. Most important was the need for each of us to understand this and not allow ourselves to get irritated by minor and temporary changes we might notice in others. This is what the best teams learn to do. I did not notice much change in my own behaviour, although I probably talked a little less, sensing that, on occasion, some of my watchmates didn't want to chat. Understanding this would turn out to be a valuable experience for me later in a wholly different environment—space.

Out of about twenty vessels that had started the race, we crossed the finish line in seventeenth position, twenty days after setting out. Although we had been beaten by the lighter, sleeker vessels, crewed by professionals, we were proud of our showing. Our untested crew had worked as a team and risen to the occasion. On a personal level, it gave me confidence that I could work with others and achieve my objectives, even in challenging conditions.

I would have the pleasure of crossing the ocean in *Pickle* a second time the following summer, in the opposite direction, from Europe back to Canada via the Azores. I would now join yet another crew in Edinburgh to bring it back to Canada. That crossing would present its own set of challenges, including being becalmed for three days on a glassy sea, shades of *The Rime of the Ancient Mariner*, when we all wondered whether we would be stranded forever, unable to find a puff of wind. Yet even that experience presented us with an unusual opportunity: to dive off the boat and swim in the middle of the ocean, with two miles of water below us. That may sound no different from jumping into your local pool, but believe me, it's a strange feeling knowing that, apart from your boat, the side of the pool is extremely far away.

In 1970, during my last year at RMC, I decided I should pursue my studies beyond the bachelor level, and so I applied for a scholarship named after Lord Athlone, the sixteenth governor general of Canada. The scholarship was awarded annually by the British Board of Trade to forty Canadian engineering students to work or study in Great Britain for two years. To my delight, I was selected. I would do my research in the Electrical Engineering Department of London's Imperial College of Science and Technology.

James Cross, the British trade commissioner to Canada, wished all the scholarship recipients farewell at a reception at Dorval airport before we embarked on a chartered flight to England. A month later, Cross was kidnapped by the Front de Libération du Quebec and would spend almost two months as their prisoner before his release on December 3, in exchange for certain FLQ members being given safe passage to Cuba. Cross had lived in a perpetual state of fear that he could be executed at any moment, and his wife and relatives had lived with the fear that they would never see him again. During this time, the FLQ executed Quebec politician Pierre Laporte, leaving his body in the trunk of a car. I learned this while I was settling into my new life in England. It truly disturbed me that a blatant act of terrorism could happen in my country; that the division between some Quebecers and the rest of Canada could escalate to the point of violence and murder.

Skipping ahead a year, the British Board of Trade sponsored a lunch for the class of '70 Athlone fellows, and the guest of honour was none other than James Cross, now back in England after his ordeal. I was seated beside him and, naturally, we engaged in conversation. I remembered him well and was shocked by the change in his appearance. I felt the life had been sucked out of him. It was not an easy conversation. I told him how much I was enjoying living in England and studying at Imperial College as we both avoided the elephant in the room, his horrific treatment at the hands of the FLQ. Let me be frank here: I love my country and my province, Quebec, but I was profoundly ashamed of what my fellow citizens did to this man and his family, and my feelings have not changed.

My time at Imperial College was a deeply creative period in my life. The university environment was conducive to this, a place where I was surrounded by other graduate students, each pursuing their own research. I believe that when you're young, you are more creative because your thinking is less constrained and more freewheeling. That's not to say that older people aren't creative, but I think we're at our best when we're young. Just as children's imaginations are not yet shackled to reality, young people's thinking is still malleable, which is a good thing, even if it leads to a lot of dead ends.

Another important consideration was my ability, without distraction, to concentrate virtually all my energy on my research, which focused on the recognition of the human face, a topic of interest to Scotland Yard, which had provided me with a grant. In simple terms, I was trying to identify the important features in facial recognition that help police reconstruct the faces of alleged criminals based on witnesses' descriptions.

We are good at recognizing someone we know, and do so almost instantly, but what are the features that make that recognition so easy? What allows our brain to process an image on the retina and quickly identify who we're looking at? Imagine for a minute somebody asking you to describe what someone's face looks like—someone you may have seen only fleetingly. What are the features that help a police artist create a likeness of that face sufficient to allow the person to be identified? Is it a specific feature like the eyes, nose, or mouth? Is it contour lines such as the hairline or the overall shape of the head? Is it the geometry between features such as eye separation or the distance between the nose and the upper lip? What is the relative importance of these and other facial features in the recognition process? That was the focus of my research.

The experience, over three years, would shape my thinking about our universities and the importance of strongly supporting them. I came to understand what a top-notch academic environment lent to higher learning, and why such institutions and spaces should never be taken for granted.

I successfully defended my Ph.D. thesis in August 1973. I was now twenty-four years old, and it was time to return to the navy, although I would no longer be living by myself. During my last year in London, I had been introduced by a mutual friend to an English woman, Jacqueline Brown, and we hit it off immediately. Like me, she possessed an adventurous streak, having gone off to Australia for a year to join friends and explore the country. She worked as an office manager in another department of Imperial College and lived in nearby Chelsea, and we began seeing each other almost every day and spending all our free time together, including on a week-long road trip to Scotland where we climbed Ben Nevis, swam in the ocean, and camped on the Isle of Mull next to the sheep. I proposed to her in the summer of 1973 and she accepted, which of course meant moving to Canada.

We married on Saturday, October 6, in St. Bartholomew's Anglican Church in Ottawa, and after a short honeymoon on the Île d'Orléans, we headed to Halifax, where we settled in to married life and I resumed my naval career. Happily for me, Jacqueline felt at home in her new country and rapidly adapted to our ways. Two years later, she would give birth at the Grace Maternity Hospital to our twins, Yves and Simone, each weighing in at over seven pounds.

Now that I had rejoined the navy, I would train as a combat systems engineer, responsible for maintaining the shipboard weapon systems and electronics such as radars, sonars, radios, electronic warfare equipment, and the command and control systems—a perfect fit for me.

This would lead to a posting in HMCS *Algonquin*, our newest Tribal-class destroyer. At the time, Canada's naval role within the NATO alliance was focused on anti-submarine warfare, and *Algonquin*, with a crew of about 250, was well equipped for the task. In addition to operating in Canada's coastal waters, we would deploy to Europe for a three-month NATO exercise. Later, I would be involved in *Algonquin*'s surface-to-air missile trials at the U.S. firing range at Roosevelt Roads Naval Station in Puerto Rico.

Serving in a warship is a unique experience. When you head out to sea, you are often on your own. You bring everything with you, only occasionally needing to replenish at sea or make a port visit. You live, work, eat, and sleep in your ship, go where you are needed, and fix whatever needs fixing. You also travel to interesting places. You witness beautiful sunsets on calm days and batten down the hatches when it gets nasty. Over time, you develop a pride in your ship and a strong solidarity with your crew. It's a special way of life.

I would also become a ship's diver, after taking an intense three-week course taught by professional navy divers where I learned to perform inspections and minor repairs under my ship. I learned to inspect the hull, rudder, propellers, and sonars and other protruding sensors.

One test during our training involved retrieving a wrench thrown ten meters from the dock into six meters of water. This was done at night, in total darkness, with no light source to assist us. The only option was to feel for the object with our hands while doing a search pattern. This was Halifax Harbour, not the Caribbean. I can't tell you how much junk has accumulated on the bottom of this harbour over the past two centuries, some of it pretty nasty. We were not to return to the dock until we had found the tool, or run out of air.

After my time in *Algonquin*, I was posted to the Canadian Forces Fleet School to teach naval weapon systems. I quickly realized that teaching requires considerable preparation. At the beginning, I was occasionally caught off guard by questions my students asked. An experienced teacher anticipates such questions. Fortunately, I had learned to admit when I didn't know the answer, telling my students I would get back to them. In other words, I didn't try to wing it!

I had learned that lesson the hard way on my second voyage in *Pickle*. Being the only one on my watch with previous ocean sailing experience, my crewmates looked up to me. One day, when we were sitting in the cockpit chatting, I said something about the transmission of radio waves that was wrong and I was politely corrected by a crewmate, in reaction to which I stuck to my position. My pig-headedness created a bit of a chill. I had allowed pride to get in the way and behaved immaturely. I later faced up to my mistake and learned an important lesson from that episode.

Teaching helped me develop some valuable skills: speaking in public, being clear and concise, and not rambling on. It also taught me to listen. Finally, it made me realize that when I spoke, my credibility was on the line. I had to know what I was talking about.

Next came a three-year stint in Ottawa, in the Department of National Defence. It was my first experience in headquarters and I enjoyed it. My work included managing the acquisition of new weapons, including surface-to-surface missiles for the navy. It was also my first exposure to the apparatus of government and how ministries functioned, a valuable experience that would prove useful in my later career in politics, particularly as a minister.

This was a period of relative calm in my life, when I rarely travelled and could be at home with my young family. It was during this time that Yves and Simone began kindergarten. Jacqueline and I bundled them up and sent them off with their backpacks, lunch boxes, and large name tags. For the first time, we were letting them out of our sight on a daily basis and for hours at a time, and we wondered how they would do. As it turned out, we needn't have worried. That's not to say that nothing ever went wrong. Like all parents, we dealt with the usual ear infections, colds, flus, bumps and scrapes, and occasionally more frightening bouts of croup, but overall, Yves and Simone's early years were relatively trouble-free.

Being avid campers, we took to the road in the summer, our Beetle struggling mightily as it towed our Coleman tent trailer, and stayed at campsites where we taught the children to swim and fish, build campfires, and connect with nature. These were perhaps the quietest and most easygoing years in our young family life.

In the summer of 1980, I was posted back to Halifax to work at the Naval Engineering Unit, providing technical support to the fleet. This included shock trials in HMCS *Iroquois*, which involved exploding large charges underwater in order to test the ship's vulnerability to the resulting pressure wave. I also participated in additional Sea Sparrow missile firing trials in Roosevelt Roads, which included a solo snorkelling expedition to find some conch shells, an outing where I came face-to-face with a shark. Fortunately, it was a young one and it quickly lost interest in me, but not before brushing its dorsal fin on my thigh, closer than you ever want a shark to get.

Although I was not particularly focused on space at the time, I remember April 12, 1981, the day NASA's space shuttle program got underway with the launch of *Columbia*. John Young was the commander and Bob Crippen, the pilot. The whole world watched as it ascended gracefully into the heavens, and we were all in awe at the thought that two days later it would land on a runway at Edwards Air Force Base in California.

While the Soviet Union had continued to send its cosmonauts into space, it had been a long time since NASA astronauts had flown, and thus the development of the shuttle had been impatiently awaited by an American public eager to see its astronauts fly again. NASA's earlier human spaceflight programs—Mercury, Gemini, Apollo, Skylab, and the Apollo-Soyuz rendezvous—had involved conventional rockets, with astronauts returning in capsules and landing in the ocean. (To this day Russia continues to use capsules, which land on the ground.) Now, for the first time ever, instead of a capsule landing in the water or on the steppes of Kazakhstan, a space vehicle would glide back to Earth and land like a plane—an almost unbelievable feat, at least as far as I was concerned.

I was taking a course at the time, and all of us paused to watch this truly historic moment on television. We all waited for *Columbia* to appear on one of NASA's long-range cameras and circle before its final approach and landing. As an engineer, I was genuinely excited, not because I was dreaming of going to space—supposedly a non-starter for a country like Canada, which didn't have an astronaut program— but because humans had designed, built, and successfully flown this extraordinary piece of technology.

In the fall of 1982, despite some reluctance on my part, I was posted to Toronto with my family to attend Canadian Forces College. Those posted there were considered candidates for higher rank, so it was not well viewed to turn down such an opportunity. My reserva- tion had to do with moving from Halifax, where our family was well ensconced in a neighbourhood we loved. I remember begging the navy to send me back the day I completed the Toronto course. Regrettably, it was not to be. Nine months later, after being promoted to the rank of commander, I was posted for the second time to National Defence Headquarters in Ottawa, this time to head up the Communications and Electronic Warfare section of the Directorate of Maritime Combat Systems.

Sadly, during our time in Toronto, Jacqueline was to experience a shattering loss in her life. Her parents called from England one night to tell her that her brother Dominic had taken his own life. Although his mental health had been fragile for much of his adult life, the news of his death still came as a shock. Jacqueline had been very close to him, as had I. It threw her into a deep depression, rendering her unable to function. I scrambled to find a psychiatrist, who recommended a stay in hospital.

Yves and Simone, who had just turned seven, could not understand what was happening to their mother, and given their age, it was not easy for me to explain it to them. Fortunately, our neighbours helped out, volunteering to watch over them after school until I returned from classes.

Over time, Jacqueline's health improved, allowing her to function once again. The news that we would be moving to Ottawa also seemed to energize her, distracting her from the tragedy she had experienced. She seemed on the way to recovery as we all climbed into our car and headed to our new home.

Returning to Ottawa meant we could again see my parents regularly. Dad had retired by then, and he and Mum still enjoyed good health. Their home was a few blocks from ours, and we would also spend time with them at their summer cottage on the Gatineau River. Dad was happy working around the house and listening to Montreal Expos games on the radio as he tended to his property. Mum was busy with her large circle of friends, reminding us frequently that although my father had retired, she did not have that option.

Meanwhile, my brothers Braun, Charles, and Philippe were getting on with their own lives. Braun had become the father of two young boys and was living in Montreal, still in search of his true calling. He would later find it when he returned to school and trained as a nurse. Charles had also married and was living in Ottawa, working as director of international affairs for the Canadian Manufacturers' Association. He would later become chief of staff to a senior Quebec minister before moving on to work in aircraft sales at Bombardier. Philippe was still single (for another year) and living in Toronto, working as a copywriter at a prominent ad firm where, perhaps coincidentally, he launched the Safari minivan with the headline "A giant leap for vankind"! He would later run his own successful advertising business.

Taking stock of my own career, I had to acknowledge that I had experienced some rocky moments early on but that my journey after that, with everything I had learned and the incredible opportunities I was given, was something for which I will always be profoundly grateful. I had entered a period of smooth sailing. That being said, life is full of surprises. Shortly after arriving in Ottawa, things would take a stunning and wholly unpredictable turn.

TWO

ON AN EVENING IN JUNE 1983, I was sitting at home after work, reading the *Ottawa Citizen*, when I stumbled on an ad from the National Research Council, or NRC, saying that Canada was, of all things, looking for astronauts. My first thought was why would Canada be looking for astronauts when we didn't even have a rocket to launch them into space? Intrigued, I learned that NASA had invited two Canadians to fly, each on a separate space shuttle mission, as a way of thanking us for having designed and built the Canadarm, the shuttle's robotic arm.

The ad appeared in newspapers across the country for several weeks and, as you can imagine, it was a much-discussed topic on talk radio and in newspaper opinion pages. Canadians, of course, were thrilled about the possibility of one of us flying in space. How could we not be?

The possibility of becoming an astronaut awakened something in me, and I couldn't brush it off. I felt myself vacillating between restrained excitement and the cold hard truth of how slim the odds were of being chosen. On top of that, I had just begun a new job, a job I loved and had worked hard to get. So why, I thought, would I want to throw a wrench in that, not to mention subject myself to the nearly sure risk of being rejected? Still, I could not shake the thought of being a pioneer on the frontier of space. I had watched the space program over the years and marvelled at the feats of astronauts and the incredible technology that took them into space and brought them back, but I had never for a second imagined myself in their place. What an extraordinary experience that would be, and not exactly an opportunity that comes along every day.

Had the path I had chosen in my life prepared me for this moment? Did I have what it took? Although I was competitive by nature and tended to grab the bull by the horns, applying to become an astronaut was so extraordinary (and perhaps absurd) that I hesitated. Fear of failure can often hold us back.

I was also conscious that Jacqueline had been through a rough patch, and as her situation appeared to be improving, I did not want to add new stress to her life. With that in mind, I approached her in a lighthearted manner, treating the ad almost like a joke. When I showed it to her, I made sure not to come across as seriously intent on applying. I asked her what she thought. She smiled as she read it and, without the slightest hesitation, told me I should apply.

When I awoke the next morning, my mind was made up. I would apply, although I promised myself I would not get upset if I was turned down, by far the most probable outcome. Jacqueline and I had planned to go camping with the twins and I was resolved to forget the whole thing for a while and get on with my life. Of course, nothing could have been further from reality. Having sent in my application, I could hardly wait for a reply and thought about it constantly.

Some time later, an envelope arrived in the mail and I opened it with trepidation. The news was good. I had made the first cut and would be considered in the next phase of the selection process. Although I did not know it at the time, over 4,300 people had applied, and the list was now down to about 2,400. Most applicants were serious candidates, while a few less qualified but undeniably enthusiastic people had also applied, including a seventy-eight-year-old grandmother! Astronaut fever was clearly gripping the country, but I tried not to allow myself to get too excited. There was still a long way to go and the odds were still much against me. At least that's how I saw it.

Having made the first cut, I now had to provide a note from my doctor. Fortunately, my health was good, with no obvious disqualifying medical conditions—including, I should add, asthma!

I also had to explain in greater detail why I felt I was a suitable candidate. In addition, the selection board asked each applicant which of two Canadian space experiments they were interested in working on if they were selected. One focused on trying to understand why some astronauts become sick when they first arrive in space, a condition known as space adaptation syndrome, while the other focused on the Canadian-developed Space Vision System, a tool designed to assist in moving payloads with great accuracy and eventually assembling the International Space Station.

Excuse the rather wooden prose, but here is what I wrote:

I am interested in the Space Vision System experiment.

Throughout my career as a Naval Combat Systems Engineer, I have been involved with the design, operation, and maintenance of complex combat systems; in particular, with radar sensors coupled to embedded processors and interactive operator displays. I am familiar with the military and non-military uses of lasers and have used them in the production of holograms. I am also familiar with raster and random scan displays, there being a large number of naval applications in current use.

My doctoral work involved the use and modification of television camera video for presentation on conventional monitors in order to conduct experiments in the field of visual perception.

I have programmed in assembler and high-level languages on several systems: PDP-11, L-304, HP 9825, and the M6800 microprocessor.

As a naval engineer, I have planned and conducted several lengthy and complex trials on naval systems. I was personally responsible for conducting the first ever anti-air defence missile firing trials aboard HMCS Algonquin. The trials covered a six-month period and culminated in the successful firing of several Sea Sparrow missiles against remotely piloted vehicles on the American Range at Roosevelt Roads, Puerto Rico. The equipment under trial consisted of a fully integrated fire control system with CW illuminator (to illuminate the target) and a slaved launcher.

In 1977, I personally designed, debugged, and set to work a small simulator for use in the training of naval weapons officers responsible for ship anti-air defence. The system is currently in use. I also participated with the Defence Research Establishment Suffield in the design of a unique Canadian naval air target for scoring of naval gunnery accuracy (the MILKCAN target).

As a naval officer, I have had to work in uncomfortable and constricted environments while discharging my duties. As a naval diver, I have also had to work under dangerous conditions. I have parachuted (as a summer student) and I have sailed twice across the Atlantic with 12 others in a 59-foot yawl. I personally feel that I am capable of working in difficult environments.

As a section head at various times throughout my career, I have been responsible for the supervision of from 12 to 37 engineers, technologists, and technicians, most of them being civilians employed with DND. I have worked with American military and civilian engineers on a large number of occasions in the past. I was a weapon systems instructor for 18 months and have addressed large audiences on many occasions.

In summary, I believe that I possess the necessary experience and correct attributes to join the team on the Space Vision System experiment. I believe you would find me fully capable of meeting the challenge which lies ahead.

Finally, I had to provide three references from people with whom I had worked. I approached naval officers who knew me well, including my current boss at National Defence Headquarters, Capt. Jim Dean. The other two, Norm Smyth and Jim Carruthers, were accomplished naval engineers who had inspired me throughout my career.

I was slightly apprehensive approaching them, not wanting anyone to think I was no longer interested in the navy. I needn't have worried. They all agreed to write letters of endorsement and wished me good luck.

Coincidentally, that summer the movie *The Right Stuff*, based on Tom Wolfe's bestselling book, was released, giving the general public a look at the process by which American astronauts were chosen and trained during NASA's Mercury program. Canadian media were more and more interested in the topic and kept asking who in Canada had the so-called right stuff. I remember talking with friends about the movie and speculating on what kind of person would apply for such a job. For obvious reasons, the consensus was that test pilots would have the inside track, although that specific qualification had not been mentioned in the NRC ad. At this point, I was keeping my cards close to my chest, telling only a select few that I had applied.

At the end of September, I received another letter from the NRC, this one informing me that I had made the next cut and that face-to-face interviews were to be conducted across the country with the remaining sixty-eight candidates. I was one of fifteen who would be interviewed in Ottawa. There would also be a security screening check and a medical exam.

The whole selection process was a bit like a lottery, where three of the six numbers on the winning ticket had been drawn and I had all three! With three to go, it was becoming more difficult for me to remain detached from what the future might hold, and how my life could change dramatically. By now, the names of the remaining candidates, including mine, had become public. Media interest was building as more and more stories featured Canadians who announced they were in the running.

The interview phase was, of course, crucial. In the process, I would also be meeting some of the other candidates. It was both inspiring and daunting to be introduced to them; they were an impressive bunch, and it helped me recalibrate my expectations. Clearly, I was competing with a highly qualified pool of candidates, which shouldn't have come as a surprise given the stakes, but now I knew.

My interview went well except for one potentially disastrous (and surreal) moment afterwards, when one of the board members, Ray Dolan, drew me aside and asked me if I drank alcohol and if so, how much. Puzzled by the question, I told him I occasionally had a beer or a glass of wine. He then produced the reference letter written by my boss, Jim Dean, and cited one sentence in the handwritten letter that appeared to state, "Marc Garneau has a regular drinking problem"! My jaw almost hit the floor and I asked if I could see the letter. Knowing first-hand that my boss's handwriting was often a challenge to read, I quickly deciphered the sentence he had written, which was: "Marc Garneau has a regular training program."

On hearing this, Ray burst out laughing. Meanwhile, I had been imagining the worst—that two illegible handwritten words could have ruled me out as a candidate. Obviously, the board must have decided to interview me on the assumption that an endorsement letter would not normally contain such a negative comment. But it certainly caught their eye enough for them to ask me about it. Despite my heart having stopped momentarily, I left the interview feeling good about it, and thankful that I had been able to clear up a potentially serious misunderstanding.

We were now in November and my family life was going well. Yves and Simone were in their own world, having made friends at school and in our Manor Park neighbourhood, oblivious that their father was seriously exploring the possibility of a second career. Meanwhile Jacqueline, whose state of mind I watched closely, was almost her old self again and seemed content with her life, our new surroundings, and her friends. That I had made it this far in the competition did not appear to be stressing her, and I was hopeful that wouldn't change, even though I dared not predict the future.

Quietly, though, I was dealing with the uncertainty of what the future held for me. My first duty, of course, was to perform my job at National Defence Headquarters, giving it my undivided attention, something I was trying to do every single day, pretending that nothing else mattered. At the same time, the tantalizing possibility that I might make it to the end of the competition and that this would change my life never left my mind. With all the hype and media attention around selection, it was clear that the chosen astronauts would become public figures, and this would carry new expectations. Of course, there was also the real possibility that I might not make the final cut, and that would feel like waking up from a dream with a disappointing ending.

In the midst of all this, thank God for my family. Jacqueline, my parents, and my brothers Braun, Charles, and Philippe all knew me better than anyone and kept me grounded. They were the anchors in my life, and as the drama of the astronaut selection unfolded, I was grateful for their calming influence.

In mid-November, I received a letter telling me I was one of nineteen finalists. Although I had prayed for this moment, I knew it was a two-edged sword: I would be either the happiest man alive if chosen or the most miserable if not. Whatever emotional detachment I had been able to preserve over the past six months was gone after receiving this good news. Now, as far as I was concerned, everything was at stake. I was all in.

The nineteen of us arrived in Ottawa in late November for a week-long assessment process that allowed the selection board to look at each of us up close. Veteran NASA astronaut Paul Weitz had also been invited to observe us.

We spent our first day at the National Defence Medical Centre, where we ran on a treadmill and were prodded, poked, and tested in every possible way to check our hearts, our lungs, our eyes, our ears, our nervous system, our digestive tract, our blood pressure, and anything else you care to imagine.

I remember one rather amusing (and slightly awkward) moment when all of us were sitting on the bus, headed to the medical centre. We were introducing ourselves to each other while each of us held a small brown paper bag containing a stool sample for testing, pretending there was nothing odd about it. Later, as we got to know each other, we would frequently laugh about that bus ride.

Another day, we each had to give a twelve-minute speech on one of four topics and then take questions. Because the subject intrigued me, I chose "The Role of Heroes." I believe society needs its heroes and role models for inspiration, particularly in difficult times. They may be ordinary people who have demonstrated extraordinary bravery. They may be athletes who have surpassed themselves. In the eyes of some, astronauts are viewed as heroes.

I gave my speech in front of a group of stony-faced board members who deliberately avoided showing any emotion while I spoke. The humour I had inserted appeared to fall flat. I left the room, feeling that I had bombed. It was a tough gig, but I suppose that was the point of the exercise.

On the question of astronauts becoming heroes, let me say this: While I've been a public figure for forty years and am used to it, the decision to become an astronaut was not a trivial one. Was I willing to be placed on an imaginary pedestal, as astronauts often are? I certainly wanted to become one, but what price was I prepared to pay in terms of public expectations? What about my privacy and that of my family? Would the public and the media agree not to cross the line between my public and private lives, or was nothing out of bounds? I accepted the public role I would later play as an astronaut but didn't know how far it might extend into areas I was not willing to share. Only time would tell, and by the time I found out, it might be too late.

One evening, we were all invited to a social event with many guests interested in speaking with us. In truth, we were being observed and assessed for our social graces. Were we comfortable in such a setting or did we feel awkward? Beyond our professional accomplishments, what kind of people were we? We were also interviewed by a national television anchor, Keith Morrison, to see how we handled the pressure of an on-camera interview.

I can't remember every detail of that week, but it certainly included many opportunities for the selection board to have a good look at each of us. It also included a visit to the David Florida Laboratory, where Canadian space hardware was tested before launch.

After that busy week of scrutiny, each of us was told to expect a call between five and seven on the following Saturday evening, letting us know whether we had made the final cut. I sat on pins and needles all day long until the phone rang around six. It was Ray Dolan and, mercifully, he skipped the niceties and got straight to the point: I had been selected as one of six Canadians who would become Canada's first astronaut corps. At that moment, a weight lifted from my shoulders, and even though I may have sounded calm, I thought I would explode. I had made it to the end of a race I thought I had only the slimmest chance of ever winning.

The NRC's decision to select six candidates surprised me given that only two flights had been promised. I would later learn that NASA also wanted to train back-ups alongside the two astronauts chosen to fly. Selecting six people would give the NRC a wider pool from which to choose.

I thought a great deal about my life that night: What if I had not seen that "Astronauts Wanted" ad? What if that BB pellet had hit me in the eye when I was ten years old? What if I had drowned in the Jacques-Cartier River when my canoe tipped over on a cold November night when I was fifteen? What if the incident that led to my arrest had spun out of control? What if my parents had not dug into my medical records and discovered I did not have asthma? What if Dr. Lavigne at CMR had not agreed to tutor me when I was floundering in chemistry? What if I had been kicked out of the navy? What if the selection board had dropped me because they thought I had a drinking problem? That evening, as I reflected on my life, I realized that fortune had smiled upon me more than once.

Although Jacqueline and I wanted to share the good news with our families and friends, we had been sworn to secrecy until the following Monday, when we would be presented to the public. My life was about to change dramatically.

THREE

EVEN TODAY, the memory of our official introduction to the public remains fresh in my mind. In addition to myself, the Canadian astronaut corps consisted of Robert Thirsk, Roberta Bondar, Ken Money, Bjarni Tryggvason, and Steve MacLean. It was December 5, 1983, and the six of us were gathered for the first time in a briefing room and told what to expect when we were trotted out for public viewing. The media would be present in large numbers, and Estelle Dorais, who handled media relations for the NRC, took firm charge of her six neophytes.

As we approached a large phalanx of reporters on our way into the event, Estelle shouted at us: "Whatever you do, don't talk to them." We followed obediently, none of us daring to respond to the journalists who, microphones extended, were trying to scrum us. The media would be free to ask us questions, but only after the NRC's president, Dr. Larkin Kerwin, had introduced us.

The buildup to this moment had lasted six months, so I was not surprised by the level of interest from the press. The place was jam-packed with reporters, and after Dr. Kerwin introduced us and explained how the astronaut corps had come into being, naturally they wanted to hear from the astronauts.

While my colleagues were dressed in civilian attire, I was wearing my military uniform. I was still an officer in the Canadian navy, even though I was now seconded to the NRC and my full-time job was astronaut. The questions directed at me focused on my military background and how my wife and two children were reacting to my becoming an astronaut. I told them that Jacqueline was quite excited about my selection but that my eight-year-old twins weren't that impressed.

None of the questions were surprising. It was clear that the assembled journalists wanted to know what kind of people we were. Were we a different breed of humans? If so, what made us different? Did we share some exceptional quality? In my opinion, not really; it was only a matter of degree.

Back then, I would have been hard-pressed to list the skills required of an astronaut because, of course, I wasn't one yet. Now, with the benefit of hindsight, I would say those skills are identical to those required in other high-performance professions. Astronauts must be able to prioritize, identifying what is most urgent at any one time; focus on the task at hand, sometimes to the exclusion of others; react correctly under stress; communicate quickly, clearly, and concisely; always keep the big picture in mind; and rebound from failure. Good motor skills and eye–hand coordination are also important, as is the ability to work in uncomfortable conditions for extended periods. Finally, and most importantly, astronauts must be team players.

With that in mind, what kind of person was I? Although I'm answering the question forty years later, I don't believe people change that much. Above all else, I'm adventurous, optimistic, and curious, and I believe that science should guide us in our decisions. I'm also willing to accept a certain level of risk in my life, what I call "controlled" risk. I have a type A personality and am driven to succeed. My mother once told me that I had the ability to focus relentlessly on a task for as long as it took. I strongly admire perseverance, hard work, honesty, and punctuality, and I appreciate accomplished people who are humble, so I tend to wince when I hear people brag. I am not charismatic or extroverted, nor would I wish to be, even if I could.

The penny dropped that morning. We were now public figures. I knew I would get used to the spotlight over time, but from now on it would become increasingly difficult for me to go anywhere without being recognized, at least locally. People would sometimes stop me in the street and ask me questions, or just say hello. Some would stare at me. I could no longer wander, anonymous and unshaven, into Canadian Tire without somebody coming up to me and wanting to chat. Starting now, I would have to be more mindful of my appearance before going out. I was stepping into the big shoes of the astronauts who had come before me, and I had an image to uphold. My cigarette-smoking habit would have to go, so in December 1983, I puffed on my last one, withdrawing cold turkey.

Although it was largely self-imposed, I felt a greater pressure to avoid personal embarrassment. I was an astronaut, and people assumed astronauts didn't make mistakes. In truth, I had not changed: I was still the same person as before. I could just as easily slip on the ice and fall on my ass. So I resolved not to take myself too seriously. After all, I was human, and like everyone, I would screw up occasionally.

It was disappointing that my extended family had to be kept in the dark about my selection and would only learn the news at the same time as the rest of the country, but the NRC, afraid of a leak, was taking no chances and kept our names secret until the last minute.

When he learned the news, my father, a man of some reserve and always under control, expressed his pride more visibly than I had ever seen before; he seemed almost giddy with the excitement of it all. He had held his emotions in check throughout the selection process, but now he let go. It's a moment I will always cherish, from a man who was not overly demonstrative. I believe he saw my success in being chosen as validation that he had done a good job raising me.

As for my mother, who had told me repeatedly when I was growing up that I could achieve whatever I set my mind to, she was also overjoyed and kept telling us that she'd known all along I would be chosen. In the days and weeks that followed, I was grateful they remained the same people I had known all my life. I needed a sense of normality, and they provided it at a time when others were treating me like a star, something I was not used to and never felt entirely comfortable with. Family matters more than we sometimes realize.

As astronauts, we were under the auspices of the National Research Council, and Dr. Karl Doetsch was appointed as our director. A glider pilot in his spare time, he had an extensive background in aerospace, including as the deputy program manager of the Canadarm project. Karl would later lead Canada's participation in the International Space Station program, and he was the founding director of the International Space University in Strasbourg, France.

Dr. Garry Lindberg, although less visible to us than Karl, also played a key role in the oversight of the astronaut program. He had been a member of the selection board that had chosen the six of us. Most notably, he had been NRC's project manager for the Canadarm, from its inception in 1974 until its successful first flight in 1981, on the second shuttle mission, STS-2. (STS stands for Space Transportation System, the space shuttle's official program name.)

It was the development of the Canadarm that made the Canadian astronaut program possible. It was designed and built in Canada by Spar Aerospace at a cost of about $108 million (under contract to NRC, with contributions from Canadian subcontractors Dynacon, CAE Electronics, DSMA Atcon, and Dilworth, Secord, Meagher and Associates) and was Canada's contribution to the U.S. space shuttle. Initially, NASA referred to it as the Shuttle Remote Manipulator System, but eventually the name Canadarm, coined by Dr. Kerwin, was adopted to acknowledge its Canadian origin.

The Canadarm would demonstrate its value on every one of the ninety missions on which it was deployed over a thirty-year period, never once faltering. It proved to be an indispensable tool for moving payloads, capturing and releasing spacecraft, helping to build the International Space Station, moving astronauts positioned on its tip as they performed tasks such as repairing the Hubble telescope, and, perhaps less known, providing important views of the orbiter with its elbow and wrist joint cameras. (In the early days of the shuttle program, it was customary to use the cameras to survey the orbiter for damage once it reached orbit.)

Ranging in age from twenty-nine (Steve MacLean) to forty-nine (Ken Money), our small cadre of astronauts came from a variety of backgrounds, bringing different skill sets and experiences to the newly constituted astronaut office. Roberta, Ken, and Bob had been selected for their knowledge and professional qualifications in the life sciences, given that Canada wanted one of our two astronaut flights to focus on how the human body adapts to weightlessness. Steve, Bjarni, and I, on the other hand, were selected because of our backgrounds in engineering and the physical sciences, given that Canada also wanted to evaluate the use of the Space Vision System on our other astronaut flight.

We set up shop in Ottawa, in Building M-60 of the NRC campus on Montreal Road. Roberta and I were paired together in one office, while Bob and Steve occupied a second and Ken and Bjarni, a third. Four people managed our training activities: Bruce Aikenhead, who ran the office; Fawzia Khan, our administrator; Parvez Kumar, responsible for our training; and Bernard Poirier, our logistics officer. In addition, because we were required to interact regularly with the media, Dr. Wally Cherwinski, the director of communications for the NRC, and Estelle Dorais both played a major role in organizing our appearances and managing our interviews. In addition, Lise Beaudoin, our unofficial den mother, organized all our trips, wherever we went in the country. We were fortunate to be supported by such a professional team.

I should make special mention of Bruce Aikenhead. He was one of the numerous Canadians who lost their jobs when John Diefenbaker cancelled the Avro Arrow program in 1959 and who were quickly scooped up by NASA as it began ramping up its astronaut program to compete with the Soviet Union. Bruce would move to the United States and play an important role in the design of the simulators used in training the Mercury program astronauts. Fortunately, he later returned to Canada and contributed to the design of the Canadarm before becoming director of the astronaut office. He was unquestionably one of Canada's unsung space pioneers. I would have the pleasure of working closely with him for many years and benefiting immensely from his expertise.

While the six of us had been chosen primarily because of our professional backgrounds, there had been no requirement for us to be test pilots, as was the case at the beginning of the Soviet and American space programs. Nevertheless, because we were astronauts, the public assumed we were all pilots of one kind or another. Whereas Ken had been a pilot in the RCAF, and Bjarni and Roberta had their pilot's licences, Bob, Steve, and I had no flying experience. I had enlisted in the navy to work in ships, not fly aircraft.

Simply put, knowing how to fly requires you to perform multiple tasks simultaneously (fly, navigate, and communicate), as well as demonstrate excellent judgment—for example, when you assess weather conditions before and during a flight. It wasn't lost on me that all these skills would be valuable for an astronaut to have, and so I began working on obtaining my licence.

Bob, Steve, Bjarni, and I purchased a used Cessna 172 and flew at the nearby Rockcliffe Flying Club, with Bjarni instructing us. I obtained my licence in 1986 and continued to fly until 2001. From the beginning, I loved it, and the first time I flew solo was an incredibly special moment, as any pilot will tell you. I regretted not having taken flying lessons earlier in my life. I had not appreciated what I had been missing. That said, you can't do everything.

As we settled into our new profession, we began studying the basics of spaceflight, using NASA-supplied class material. We also learned about the space shuttle and its three main parts. First is the orbiter, which is the winged vehicle in which the astronauts live and work during their mission. Second is the external tank, which is the fuel reservoir for the liquid hydrogen and oxygen that power the orbiter's three main engines. Once the main engines are turned off, the external tank separates from the orbiter and re-enters Earth's atmosphere. Lastly there are the two solid rocket boosters, which deliver millions of pounds of thrust during the first two minutes of flight, before they are jettisoned and later recovered at sea for reuse.

We also focused on the principles of orbital mechanics and the physics of getting to space, orbiting Earth, and returning in a controlled manner. There was a lot of ground to cover, but we could only go so far in Canada. At some point we would need to travel to the Johnson Space Center in Houston, Texas, the NASA facility responsible for astronaut training, where we could train in high-fidelity simulators.

While we were learning about spaceflight, we were also introduced to the experiments we would perform on orbit. One of them, led by Dr. Doug Watt of McGill University, would examine whether the human body, which has evolved in gravity, would be affected by weightlessness. In addition to Dr. Watt's expertise, we were fortunate to have Ken Money, a physiologist, in our corps. Ken was an internationally recognized expert in vestibular physiology, and we learned a great deal from him, including how the vestibular system, which helps us with balance and spatial orientation, is affected by weightlessness. Some astronauts experience symptoms of motion sickness, including nausea and vomiting, during the first few days of spaceflight, a potentially problematic situation if there's an emergency and you're unable to act at full capacity.

By the way, when I speak of weightlessness, I don't mean that the force of gravity is not acting on us. When we are orbiting a few hundred kilometres above Earth, that force is not that much less than on Earth. The result is that both the astronauts and their orbiter are constantly being pulled back towards Earth while also moving very quickly forwards, which is what causes our curved trajectory around the planet. In effect, the astronauts and their vehicle are constantly free-falling, and the net effect is that the astronauts float relative to their vehicle. Not wanting to overwhelm their audience with the physics involved, astronauts sometimes resort to the shorthand of saying there is no gravity in space, when what they really mean is that we are constantly free-falling.

As part of our training, it was decided that the six of us should become intimately familiar with our motion sickness symptoms (something already known to me, having experienced seasickness during my naval career). This meant being strapped into diabolical machines that would spin or tumble us with the express purpose of trying to make us sick. We were subjected to this on many occasions at the Defence and Civil Institute of Environmental Medicine in Downsview, Ontario, and I turned out to be the most susceptible of the group! I came to dread my trips to Downsview, experiencing a Pavlovian bout of nausea each time I approached the institute for a torture session. That said, I vowed to be stoic about it, and I made a point of never asking the machine operator to stop spinning or tumbling me until I reached for my bag and vomited. I am not ashamed to say that I filled many bags, although I can also say that over time my tolerance increased.

You might naturally assume that a person who experiences motion sickness on Earth will also experience it in space, but the scientific evidence at the time did not indicate a strong correlation. Some astronauts who were seemingly "bulletproof" to motion sickness on Earth succumbed to it in space, and vice versa. To my knowledge, this phenomenon remains unexplained to this day.

While our training was going on, Canada was faced with an important decision: Who would be the first Canadian to fly? NASA had issued an invitation for a Canadian to train as a payload specialist, a special category of astronaut, and if successful, that person would fly on the space shuttle in the fall of 1984. The deadline for Canada to make its selection was March 1, less than three months after the six of us had been named. Amazingly, this meant a Canadian would fly less than a year after being chosen, a demanding though manageable timeline.

As a payload specialist, the chosen astronaut would be responsible for conducting five Canadian experiments. Unlike the professional

astronauts, known as pilots and mission specialists, payload specialists are regarded as invited passengers, albeit with duties to perform.
They generally carry out scientific investigations while the other
astronauts do everything else. That said, payload specialists still have
to undergo a certain amount of training. You can't just hop on the
shuttle like an airline passenger and do your experiments. Training is
required to ensure you won't be a liability, and once in space, you're
expected to perform a number of shared tasks such as preparing
meals and taking photographs, tasks that contribute to the mission
and, more importantly, integrate you into the team.

Mission specialists, on the other hand, are career astronauts
with much broader responsibilities. These might include operating
the Canadarm, or doing an EVA, or extravehicular activity, the NASA
term for a spacewalk. A mission specialist is required to know all
the shuttle's systems in detail and will support the pilot and commander during critical phases of flight, such as launch and re-entry
and on-orbit rendezvous and docking operations. On missions to
the International Space Station, there are even more systems to
learn. Eventually, Canadian astronauts would train as mission specialists, but that was still some time in the future.

Pilot astronauts are the ones who fly the space shuttle when it is in
manual control and who eventually become mission commanders.
They are all former test pilots. Because the space shuttle belonged to
NASA, only U.S. citizens were eligible to become pilot astronauts and
eventually mission commanders, a policy maintained throughout the
life of the program.

On the first day of March 1984, the director of the Canadian astronaut program, Dr. Karl Doetsch, convened a special meeting of the
astronauts. We all gathered in a small conference room, and while we
didn't know for sure, we suspected why the meeting had been called:
Karl was going to announce the name of the first Canadian to fly.
I felt all six of us were qualified, so, as far as I was concerned, I had a
one-in-six chance.

Karl walked into the room and, without beating around the bush, announced that I would be the first to fly and that Bob Thirsk would be my backup. Although I'd been hoping for this moment, it still caught me by surprise. Yes, I was ecstatic, but I was also conscious that an enormous responsibility had now been placed on my shoulders. Since no Canadian had ever been to space, my country would be watching me with heightened expectations, and I would have to rise to the occasion, a sobering realization. My performance would also set the tone for future Canadian space missions, so a lot was riding on how well I did.

If any of my colleagues were disappointed not to have been selected—I surely would have been—they hid their feelings well and were gracious about it. On the one hand, it meant they would wait a little longer for their opportunity, but on the other, it was comforting to know that the first flight of a Canadian was imminent with, hopefully, more flights to come. Everyone congratulated Bob and me, and in turn we thanked them for their show of confidence. I told everyone I would need their support and made a promise that I would do my best to make them proud, a promise I did not take lightly.

With my heart pounding, I went home and told Jacqueline, who was overjoyed. As we both absorbed this life-changing news and all its implications, we realized it would draw more attention to our family, which is exactly what happened.

The next day, Bob and I joined the minister of state for science and technology, Donald Johnston, as he announced our names at a press conference. The first question from the media was predictable: "Why do you think you were chosen to be the first to fly?" The truth is, I didn't know. There was some speculation that my military background gave me an advantage, but I don't personally believe that was a factor. Was it my prior experience of sailing across the ocean, demonstrating that I could work as part of a team in a difficult and demanding environment? Perhaps. Was it the fact that I was bilingual? It's certainly possible that the Canadian government wanted the first astronaut to fly to be proficient in both official languages. Having said that, nobody involved in my selection ever confirmed any of this to me. Perhaps it was for all those reasons, or maybe none of them. In the end, it didn't matter. What mattered was the job ahead of me.

Bob Thirsk also received many questions. One was: "How do you think you will feel, knowing that you will undergo the same training as Marc but that in the end, you will probably not get to fly?" I loved Bob's answer. He said it would be like going to an incredible movie and having to leave ten minutes before the end! Having said that, Bob also made it clear that he was happy to be my backup and that he would be there to support me. He was true to his word. Bob would eventually fly to space twice and spend almost 205 days on orbit. He has remained a dear friend, largely because of how we worked so well together when I was preparing for my first flight.

After the official announcement, the media descended upon our home and wanted to know what our children thought of their father's new job. While Yves and Simone understood that I was going to fly on the space shuttle, they could not yet grasp its significance. They were barely eight years old.

Media interest was intense, and nothing seemed to be off limits. One local TV station brought its cameras, lights, and microphones into our children's bedroom while Jacqueline and I read them bedtime stories. Another filmed them at their lemonade stand in front of the house and asked them questions. Fortunately, they didn't seem bothered by all the attention. Naturally, there was also a great deal of interest in Jacqueline, who would play an important role in supporting me and the children in the months ahead. She took it all in stride.

Transitioning from being a private citizen to becoming a public figure is not a trivial adjustment. Many have found it difficult, sometimes with unfortunate consequences. It requires a certain discipline, not to mention maturity, and knowing when to draw the line. In adapting to my new status, my goal was to preserve, as much as possible, a normal routine in my life and that of my family. I didn't want

Jacqueline or the children to be changed by all the attention being lavished on them. (I had the same concern for myself.) On rare occasions, I drew the line on certain questions being asked of me or of requests to follow my family around as they went about their lives. I believe we all need to feel that some things can remain private.

Not long after I was chosen, the mission on which I would fly was announced: STS-41G aboard space shuttle *Challenger*, an eight-day mission scheduled for liftoff on October 5, seven months away.

The curious system for numbering flights was partly based on superstition. For the first nine shuttle missions, consecutive numbers were used, but NASA wanted to avoid using the number 13. The *4* in *41G* referred to fiscal year 1984—which for NASA began on October 1, 1983—while the *1* designated a launch from the Kennedy Space Center at Cape Canaveral (*2* was reserved for future shuttle flights from Vandenberg Air Force Base in California, which never occurred). The letter at the end, in my case *G*, the seventh letter of the alphabet, identified the seventh flight of the year. Needless to say, this numbering system was flawed for more than one reason. For example, my flight, originally scheduled to occur in late September, slipped into October. Based on my actual flight date, I was on the first flight of NASA fiscal year 1985, which normally would have received the designation 51A. Ironically, my flight turned out to be the thirteenth flight of the space shuttle!

For the sake of historical accuracy, I should mention that I was originally assigned to a different flight, scheduled to fly around the same time. But before I had a chance to meet the crew, I was reassigned to STS-41G. This speaks to a certain fluidity in flight assignments in those early days when NASA was planning many flights in parallel and trying to figure out where best to slot me in.

As I continued preparing for my flight, I was getting blitzed by media requests. How was my training progressing? How was my life changing? What would my life be like after I returned to Earth? Was

I nervous about it all? Did I realize that I would be making history as the first Canadian astronaut? When would I be going down to the Johnson Space Center to meet my fellow crew members and train with them? I remember in particular Ellie Tesher, a columnist for the *Toronto Star*, who wanted me to bare my soul every time we met.

Although I understood the curiosity that Canadians had about me as the first to fly, I had only just started training for my mission and wished I could escape the intense media scrutiny until *after* I had flown and proved myself. I think I was trying to ignore the pressure that comes from high expectations. The NRC had warned me this would happen but urged me to accept the more important requests. I acknowledged that making myself available for interviews was part of the job. It would not be good if Canadians saw me as aloof and remote. In addition to giving me a great deal of positive public exposure, the media were also building public awareness of what would be a historic Canadian event, even though, on occasion, it wasn't obvious. Sometimes there was too much focus on me and my family instead of on the significance of Canada's first participation in the human exploration of space. Nevertheless, I resolved to make myself as accessible as I could while I was still in Ottawa, since this would become difficult once I relocated to Houston, where there were no Canadian journalists and NASA would control my schedule.

Bob and I were told to report to the Johnson Space Center for training at the beginning of August. We both rented apartments in Camino Village near JSC and settled in. I was on my own now, my family having remained in Ottawa. We were eager to begin our training, although, as payload specialists, I noticed that some of the American astronauts welcomed us with less than open arms, as if to say that we were not real astronauts. There may also have been a feeling among some that we were stealing their jobs, which was understandable. Some of them had been training for years, hoping for a chance to fly, and here I was, a Johnny-come-lately, from another country no less, who was getting a seat on the space shuttle less than a year after being selected.

Although none of this was ever expressed publicly, and I never spoke about it, it was palpable, and part of my reality at JSC. As if to underscore the difference between payload specialists and the other astronauts, Bob and I were assigned an office in a different building from that of our crew. Normally, all crew members assigned to a mission work out of a single office so they can build team cohesion while preparing for their flight. Despite the obvious benefits, Bob and I were invited to join our crew only when we met in a simulator session that required everyone's presence. Even Paul Scully-Power, an American payload specialist who was also assigned to STS-41G, had his office in a different building.

I should point out that payload specialists came from a variety of places and backgrounds: they could be foreign astronauts like myself or Ulf Merbold, a German astronaut from the European Space Agency, or Patrick Beaudry from the French Space Agency. They could be American scientists or engineers with special skills, like Byron Lichtenberg and Charlie Walker. They could even be American politicians like Senator Jake Garn or Congressman Bill Nelson, who would later become a NASA administrator. Over time, thirty-six shuttle flights would carry sixty payload specialists into space (some, more than once), with Paul Scully-Power and me becoming the fourth and fifth to fly. After Ulf Merbold, I would be the second non-American to fly on a shuttle mission.

Fortunately, the practice of segregating payload specialists would end shortly after my flight, based on my recommendation that they should share the same office as their fellow crew members. Notwithstanding that they were not professional astronauts, they were part of a crew and should not be, or feel, isolated. Building team cohesion was paramount and had to trump any ill feelings related to professional status. It was a significant cultural shift for NASA's Astronaut Office, but one that I believe was necessary.

My training at JSC involved a variety of tasks. Most importantly, I had to acquaint myself with different emergency scenarios, such as a loss of cabin pressure or a fire, and the procedures to follow on the launch pad, on orbit, and after landing. For example, if evacuating the shuttle was necessary during countdown, we would exit the vehicle, jump into baskets, slide down a long cable to an arresting net on the ground, and then sprint to a nearby bunker. If we landed somewhere other than intended and could not exit the orbiter through the side hatch door, there was a procedure to exit the vehicle through the overhead flight deck windows, allowing us to slide down to the ground using ropes.

I spent a lot of time learning about the orbiter, the vehicle that would become my home for eight days. This included how to use the communications equipment to speak with Mission Control. And of course, weightlessness can present challenges. In fact, very little is straightforward when you're weightless. A case in point is the treadmill used for exercising. Without bungees to hold you down while you're running, you simply float away! Even routine tasks that I had long ago mastered on Earth required special training, whether it was preparing and eating food without making a mess, performing daily ablutions, or going to the bathroom using a specially designed piece of equipment known as the Waste Management System, in which suction replaced gravity to draw away your expelled but otherwise floating urine or fecal waste.

Photography lessons were also part of the curriculum. Every astronaut was expected to record as much of the mission as possible, whether it was crewmates working inside the vehicle or views of Earth. This meant learning how to use many photo, film, and video cameras and being briefed on various terrestrial sites of interest.

"Spacewalking," of course, is a misnomer. In space, whether inside or outside, you don't walk, you float, and to get around, you use your arms or legs to apply a force against a surface in order to move you effortlessly from one place to the next. Moving gracefully is an acquired skill. To learn what it's like, we boarded a special KC-135 reduced-gravity aircraft that flew parabolas in the sky. Each parabola would allow us to experience roughly twenty seconds of weightlessness.

The KC-135, the size of a Boeing 707, would typically fly forty parabolas on each sortie. A large aircraft climbing and diving repeatedly is an impressive sight to behold. The plane climbs at an angle of forty-five degrees, arcs over, and dives at an angle of forty-five degrees, then pulls up and repeats the same manoeuvre for each successive parabola. At the top portion of the arc, the passengers begin to float upwards, a sensation that is both wonderful and strange. When the aircraft pulls out of its downward trajectory, the passengers experience a 1.8 g force that pulls them down forcefully to the floor of the aircraft, making them feel almost twice as heavy as usual. This repeated cycling between floating and feeling heavy can play havoc with your vestibular system and can lead to varying degrees of motion sickness, which is the reason the KC-135 was known as the "Vomit Comet"!

The first time I flew in it, I became sick on the seventh parabola and on each subsequent one (they flew forty-five of them that day). It was worse than the seasickness I had experienced during my naval career and it took me almost a full day to recover. Despite how awful I felt after that first experience, Bob Crippen, my mission commander, urged me to keep flying until my body got used to it. I followed his advice and climbed aboard the KC-135 again and again until I could complete an entire set of forty parabolas without experiencing any symptoms. In the end, I was even enjoying myself. It was much like getting my sea legs!

The true test, proving that you had conquered the parabolas, was heading over to Pe-Te's Cajun Bar-B-Q House for lunch after a sortie and eating a pulled pork sandwich smothered in sauce, a long-standing tradition among astronauts and one I enjoyed for many years.

In addition to my payload specialist training, I had to get up to speed on the five Canadian experiments I would be conducting on orbit. This meant working with five scientific teams, first in Canada and later at JSC, and familiarizing myself with the procedures for each experiment.

Known collectively as CANEX-1, the five experiments would focus on three areas: space technology, space science, and life science. In the space technology experiment, I would record video from cameras in the payload bay of targets on the underside of the Earth Radiation Budget Satellite as it was being manoeuvred with the Canadarm by Sally Ride and subsequently released to space. This video would be analyzed by NRC to assist in the development of the Space Vision System, a tool that would later be used to help build the International Space Station.

The space science experiments focused on the physical properties of the space environment and of the earth's upper atmosphere. While we often refer to the vacuum of space, it is not a perfect vacuum. Atoms and molecules exist in small concentrations in the upper reaches of the atmosphere and can impact spacecraft and space vehicles in different ways. In one of the experiments, called SPEAM (Sun Photometer Earth Atmospheric Measurements), I would point a device called a photometer directly at the sun as it passed though the atmosphere during sunrises and sunsets to measure atmospheric constituents, including water vapour and pollution. Of particular importance was gaining an understanding of how atmospheric gases affect the chemistry of the ozone layer.

A second experiment, called ACOMEX (Advanced Composite Materials Experiment) would examine the effects of the space environment on certain non-metallic composite materials used in the manufacture of spacecraft. The materials would be taped to the Canadarm and exposed to space in the direction of flight for maximum impact. I would photograph them and report on whether I noticed any visible degradation in the samples.

In a third science experiment, OGLOW (Measurement of Optical Emissions [Glow] on and from the shuttle orbiter), I would photograph the orbiter's vertical tail at night, using a camera fitted with special lenses and an image intensifier (similar to those used in night goggles) to see whether a reddish glow was detectable as a result of atomic constituents colliding with the tail. Any glow produced from this type of interaction could be detrimental to certain optical instruments used in space.

My last experiment, SASSE (Space Adaptation Syndrome Supplemental Experiments), would focus on possible changes in the human body in weightlessness. Because we have evolved in gravity, can we expect to experience taste, smell, and touch differently in space? Will our vestibular system, which provides the brain with important information about body orientation and helps with balance, behave differently? Would I, a person susceptible to motion sickness on Earth, experience the same symptoms in space?

Finally, other than my own experiments, I was also expected to be knowledgeable about the main objectives of mission STS-41G to be carried out by the rest of the crew: Earth observation using a large radar and various cameras located in the payload bay, the deployment of the Earth Radiation Budget Satellite, a spacewalk to test an orbital refuelling system, photographing key ocean features, and shooting IMAX footage for the film *The Dream Is Alive*.

The crew of STS-41G consisted of the following astronauts: Mission Commander Bob Crippen, a veteran of three flights; Pilot Jon McBride, on his first flight; Mission Specialist 1 Kathryn Sullivan, on her first flight; Mission Specialist 2 Sally Ride, on her second flight; Mission Specialist 3 David Leestma, on his first flight; Payload Specialist 1 Paul Scully-Power, on his first flight; and Payload Specialist 2, myself. I was the only non-American in the crew, although Paul was born in Australia and subsequently became a U.S. citizen.

In addition to Kathryn Sullivan becoming the first American woman to do a spacewalk, our mission would include two other firsts: it would be the first time a crew of more than six would fly on a space mission and the first with two women.

Normally, crews train for up to a year before they fly. During that time, they get to know each other well and form strong ties. They trust each other, know each other's strengths and idiosyncrasies, and know exactly how and when to back each other up. Somehow, Bob Thirsk and I had to integrate ourselves into this tightly knit group barely two months before launch. Given that I was going to be living and working with six other people I had only a short time to get to know, being accepted was vital. Would the crew welcome me as one of their own?

As it turned out, I need not have worried—I joined a fabulous crew. From the beginning, Bob Crippen, known to everyone as "Crip", took me under his wing and made me feel comfortable and accepted. He was already a veteran of three shuttle flights, including the first one in 1981 with John Young, which I had watched with great interest when I lived in Halifax. He radiated confidence and a high level of professionalism, and I looked up to him. Jon was an affable and easygoing West Virginian who also made me feel I was part of the team. Kathryn and I shared a Canadian connection—Halifax. She had conducted oceanographic research at Dalhousie University when I was serving in the navy. Sally and Dave were more aloof before the flight, but all that would change on orbit. In fact, I worked closely with Sally during one of my experiments, as she put the Canadarm through a series of manoeuvres while I recorded the motions using a payload bay camera. Finally, Paul and I shared something in common—we were both payload specialists. He even agreed to be a subject for one of my experiments.

STS-41G would be *Challenger*'s sixth flight, with its primary focus being on Earth observation, at an orbital inclination of fifty-seven degrees. This was good news for me because it meant we would regularly be passing over much of Canada (except for the Territories). Since the shuttle would be flying upside down, the views of Earth from the overhead windows on the flight deck would be spectacular.

As the launch date drew near, media interest continued to build, and most Canadian news outlets planned to send their reporters to Kennedy Space Center for the launch and then on to the Johnson Space Center to follow the on-orbit phase of the mission. To my surprise, a small weekly from Bowen Island, B.C., called the *Undercurrent* would opt instead to show up for the landing to interview me on the tarmac, something no other Canadian news outlet chose to do. My youngest brother, Philippe, now twenty-five years old and living in Toronto, also wanted to get in on the action. He persuaded the Montreal radio station CKAC to hire him for daily reports from Houston throughout the mission, so he could add a personal dimension to the coverage, an attractive proposition for any media outlet. Kudos to him for taking such an initiative.

It was an unwritten rule at NASA that crews should be isolated from all non-mission-related activities in the three weeks before launch so they could fully concentrate on their upcoming flight. This rule, considered non-negotiable, would be broken in our case. Brian Mulroney had just been elected prime minister and was eager to meet President Ronald Reagan as soon as possible. Accordingly, Reagan's office arranged a meeting for Mulroney at the White House and it was decided that I should be present, along with two of my fellow crew members, to demonstrate the strong ties between our countries. This news was not received enthusiastically at the Johnson Space Center, but what can you do when the president of the United States makes a request? Only God and POTUS could overrule NASA! And so, Bob Crippen, Kathryn Sullivan, and I made our way to Washington, D.C., and reported to the White House, where we were ushered into the Oval Office to meet the president and the prime minister.

Without resorting to overstatement, I was a witness to history that day. Being in the Oval Office, with my prime minister and the president of the United States, as they met each other for the first time, was an unforgettable moment. It was a critically important encounter, not only for the prime minister but also for the future of Canada–U.S. relations, the chance to cement a friendship between neighbours. The benefits were incalculable. It was clear that the two leaders hit it off from the start, perhaps because of their common Irish ancestry but also, no doubt, because of their conservative politics. Their conversation was easy-going and light-hearted. This first encounter would set the stage for a productive relationship throughout their overlapping terms, and would eventually lead to the North American Free Trade Agreement. The lesson was obvious: good personal relations with our southern neighbour, starting at the top, make a big difference. It has not always been the case.

Following the meeting in the Oval Office, the president and prime minister moved to a different room for a press conference, accompanied by Crip, Kathryn, and myself. I will always remember the president, with a smile on his face, showing us a little note his press secretary had placed on his podium, warning him that the microphone was live. This was because of his earlier statement, during a weekly radio address from his ranch in California, when, in reference to the Soviet Union, and not knowing his microphone was live, he had said: "We begin bombing in five minutes"!

Shortly after my flight, I would be asked to appear at a gala event in Quebec City where the prime minister and the president met for the so-called Shamrock Summit and famously sang "When Irish Eyes Are Smiling." My job, as instructed by someone in the Prime Minister's Office, was to rise through a trap door on the stage wearing my flight suit, with spotlights and artificial smoke swirling around me, look at the audience, and pronounce the immortal words: "Take me to your leader." At the time, I had more than mixed feelings about this, suspecting that the idea had been hatched by a nineteen-year-old in short pants. But the audience ate it up, which gave me some relief, considering how hokey it was. My reluctance had nothing to do with the prime minister, whom I liked, but rather with the fact that I was not a natural extrovert, causing me to feel slightly uncomfortable. I returned to my hotel that night and swore to myself: never again! Deciding on my own to ham it up is one thing, something I would do on many occasions as transport minister when it was time to clear Santa Claus for his Christmas flight in Canadian airspace. But being instructed to do so by someone in the PMO was something I did not appreciate at the time.

In the final week before launch, the crew goes into quarantine in specially designed quarters at JSC. The main reason, of course, is to limit exposure to germs, but it's also an important transition point. The days-long countdown has begun, and instead of going home to your family after work, you move in with your crew, acutely aware that your launch is fast approaching. (The transition was easier for me than for others since my family was back in Canada.) In some ways, it feels like you're retiring from the world and living in a cocoon, making contact only with those who truly need to see you. You are disconnecting from the comings and goings and trivialities of everyday life, preparing for an otherworldly journey.

While in quarantine, the crew reviews what they will do on orbit, and additional simulator sessions are scheduled to keep everyone sharp. The crew also uses this time to shift their sleep cycle in order to get a full night's rest before the scheduled wake-up time on launch day, which in our case was at 2:45 a.m. The rooms used for sleeping are windowless and therefore pitch-black when the lights are off. Other rooms where you spend your "awake" time, which may be in the middle of the night, have their ceilings completely covered with fluorescent lights. Having pitch-black rooms for sleeping and well-lit rooms for awake time helps to shift your biological clock.

The segregation in crew quarters is also meant to prepare you mentally. It is a time of reflection as you think about what lies ahead. It is a time to concentrate all your attention on one critical task— your mission.

Three days before launch, the crew flies to the Kennedy Space Center, or KSC, where similar crew quarters are provided. On arrival,

I was asked what I wanted for my "last meal," which would be breakfast on launch day. As a bit of a joke, I asked for kippers, which drew a blank look from the kitchen staff. I had developed a taste for them when I lived in England, but it was clear that Floridians were not familiar with this type of smoked fish. To their credit, the kitchen staff scoured all of Florida and found my kippers. In the end, I ate them the day before launch and chose a more traditional breakfast on launch day.

The last three days before a flight are designed to help you to relax and wind down after a long period of training. KSC has a private beach house where the crew and certain family members can get together for a barbecue and enjoy each other's company one last time before launch. Although our children could not join us, I was able to invite Jacqueline and my parents, after they had been cleared by the doctor.

It was a lovely day, and I took a long walk on the beach with Jacqueline. We talked about the flight itself and what would happen after landing. We both focused on the positive. Yes, we knew the flight carried some risk, but we chose not to dwell on it. In that respect, I suspect we were no different than other astronaut families. My children were only eight years old at the time, so they didn't understand the risk. Jacqueline, my parents, and my brothers, on the other hand, did, but they saw no need to talk about it. While I sensibly got my affairs in order, we all chose to focus on the best possible outcome.

On the night before launch, our crew went to bed at seven. We were scheduled to lift off at 7:03 the next morning, but we'd have to be up before three. Because of the excitement ahead of me, I was in no mood to sleep, but I knew a good night's rest was imperative. Equally important, I wanted to prove to myself that I had everything under control and, despite the excitement, that I could clear my head and fall asleep.

I felt like a child the night before Christmas, although what I was about to experience was far more complex than excitement. It would be the beginning of an extraordinary personal journey—a journey that few humans had ever undertaken. Of those on our mission, only Bob Crippen and Sally Ride had flown before. For the rest of us, it would be a totally new experience. We would become the 148th to the 152nd humans to fly in space, and in my case, I would be the first Canadian. There was a profound recognition on my part that I was a privileged human being.

I fell asleep in about forty-five minutes and was awakened at 2:45 as planned, feeling well rested. I shaved, got dressed, and reported to the flight surgeon for a final checkup. My vitals were excellent. I then joined the crew for that "last meal," a public event with NASA cameras rolling. The truth is that most astronauts don't eat or drink much just before they fly; I'll leave it to the reader to guess why.

After breakfast, I met briefly with Tom Siddon, the Canadian minister of state for science and technology, who had flown down to wish me good luck on behalf of Canada. I was touched by this gesture. It also reminded me that a good number of Canadians would be turning on their television sets that morning.

My flight aboard space shuttle *Challenger* would be its sixth. *Challenger* would launch on four more occasions over the next fifteen months before the tragic disaster in January 1986, when the entire crew was lost. That catastrophic event would lead to many changes, such as dressing astronauts in bulky orange flight suits known as Launch and Entry Suits that could be pressurized if there was a cabin leak. Because my flight occurred before any of these changes were made, crews still wore light-duty flight clothing, essentially a one-piece pilot flight suit which could not be pressurized, although our helmets could be sealed. (On orbit we changed into pants or shorts and polo shirts.)

Before we went to the launch pad, a suit technician checked my helmet to ensure there was a good seal and a working headset for communicating. Then, at the request of JSC, some electrodes were connected to my chest to monitor my heart rate during launch. I didn't have any concerns with NASA keeping an eye on my vitals, nor did Paul Scully-Power. However, my guess is that the rest of the crew did not agree to being "instrumented." Astronauts are reluctant to have their vitals monitored in case something unexpected is discovered, preventing them from flying, at least temporarily.

We thanked the crew quarters staff, took the elevator to ground level, and headed out to the "Astrovan." A horde of media had assembled behind a cordon and were filming us as we walked by and waved. I remember Eve Savory from the CBC shouting two questions at me as soon as she saw me: "How did you sleep, Marc?" and "How are you feeling?" To which I replied that I had slept well and was feeling great. We boarded the Astrovan and drove to the launch pad. It was still the middle of the night as we approached *Challenger*, lit up against the night sky, with the countdown continuing under the watchful eye of KSC Launch Control. It was a breathtaking sight.

For safety reasons, only a small number of people accompany the astronauts to the launch pad. Their job is to secure us in our seats and then close the side hatch door and depart. When we arrived, we took the elevator to level 195 and, one after the other, we boarded the vehicle and were strapped in, with critical items such as flight procedure books placed within easy reach. I was seated in the mid-deck, the compartment below the flight deck, and faced a wall of lockers, with Paul Scully-Power right beside me. Dave Leestma was seated in another part of the mid-deck, and the other four were on the flight deck. When all seven of us were secured in our seats, the side hatch door was closed. The countdown clock was at roughly two and a half hours, and apart from us, no one else remained on the launch pad.

To call this the longest two and a half hours in my life is an understatement—it felt like an eternity. Fortunately, I could follow everything that was going on. Most of the voice traffic was from Launch Control, as fuelling of the large external tank continued and each shuttle system was checked and rechecked in preparation for launch. The weather report was also being updated, both at KSC and at the transatlantic abort sites where the shuttle might have to land if something went wrong. There was little idle chatter from the crew. I suspect that, like me, they were reflecting on personal matters, although the astronauts on the flight deck were also going over their procedures one last time before liftoff.

It was a time of deep introspection for me, one of those moments when something dramatic is about to happen and you must be ready for it. You can't delay what's coming and it's too late to change your mind. In a little over two hours, I would be launching into space. I had worked hard to reach this moment, but now that it had arrived, was I ready for it? Had I trained hard enough, worked out the kinks, and prepared myself for every eventuality?

I thought about Jacqueline, Yves, and Simone watching from the roof of the Launch Control building, surrounded by other astronauts ready to support them if they needed it. Had I told them how much I loved them? Had I left anything unsaid? I needed to feel there would be no regrets, whatever happened—that my mind was at peace. I visualized them with the other families, waiting for the launch. I hoped they were excited but also relaxed. I had brought along a picture of them in my personal notebook that I would look at often, anchoring me back to Earth.

I also thought about my parents and friends watching from another viewing site. They would not benefit from astronaut support if anything went wrong (that would change after the *Challenger* disaster). I hoped someone was there to answer their questions if there was a problem or a delay. I knew that media cameras were pointed at them to catch their every emotion before and during liftoff. I knew that my father, who had often experienced the fear of death during the war, was probably handling it well, and that my mother was pretending to be relaxed but was probably a nervous wreck. I was grateful to have a father and mother who had played such an important role in getting me to this point in my life, despite the many challenges I had presented them with along the way.

I thought about my fellow Canadians too. I had not been back to Canada in over two months, but I knew they were watching, and I wanted them to feel proud of this moment. My country had given me so much. It was time for me to deliver.

As I awaited liftoff, a few butterflies fluttered in my stomach. The truth is that any normal human being would feel this way before the release of seven million pounds of rocket thrust and a critical sequence of events that will take you from a standstill to Mach 25—twenty-five times the speed of sound, or roughly 28,000 kilometres per hour—in eight and a half minutes. So yes, I was a little nervous, but certainly not so nervous that I wanted to get out of the vehicle. What lay ahead was too irresistible to miss.

Fear is a complex emotion. Some people are fearless while others are scared of almost everything. Neither extreme is normal. On the positive side, fear can prevent you from doing something that is unnecessarily dangerous or downright stupid. Fear is about assessing risk. I try to approach this in a rational and logical manner. I do it whether I'm skydiving, flying a plane, scuba diving, or doing any activity where there is an element of danger. I know there are risks involved in the mundane tasks we perform every day, such as crossing the street or slicing bread with a sharp knife, but the risks are so small that, after taking precautions, I won't hesitate to do them. On the other hand, launching into space involves greater risk with potentially greater consequences. I thought about it seriously beforehand and asked myself, Is it too much of a risk? The answer was no. I understood the science and technology of spaceflight and trusted my crew and those on the ground who had prepared the vehicle for flight and who would be monitoring every second of our mission.

In my life, I have had moments of doubt about my faith. That morning, I said a prayer, a prayer that God existed.

As the countdown proceeded without any hitches, I became more and more confident we would be launching that morning. The sun, still below the horizon, was beginning to light up the eastern sky and the weather looked good, with only a few scattered clouds above the launch pad. All systems were "nominal" (NASA jargon for behaving normally) and the shuttle was beginning to come alive. There is always a planned hold in the countdown at T–9 minutes (T stands for the exact time the shuttle is scheduled to launch) so that the launch director can get a final "go for launch" from all the members of the launch team. When all systems were given the green light and the countdown resumed, I knew we would launch.

The long wait was nearly over, and the pace was quickening as I listened to the voice traffic between Launch Control and the crew. It was time for me to focus on what lay ahead. I would not allow extraneous thoughts to intrude from this moment onwards. Although I had no role to perform during this stage, I knew exactly what needed to happen and in what order, and nothing else mattered. It was too late for anything else. I was now totally committed.

At T–5 minutes the vehicle's three auxiliary power units were activated to deliver hydraulic power to the orbiter. This was a crucial step, and all systems continued to look good as we approached the final ten seconds before launch. At T–6 seconds, the first main engine was powered up, followed by the second at T–4 seconds and the third at T–2 seconds. I could hear and feel the roar beneath me, and at T=0, the two massive solid rocket boosters on each side of the external tank ignited and we lifted off the pad. Millions of pounds of thrust had been released and nothing could hold us back. We were headed for space, and it would be the most extraordinary ride of my life.

FOUR

IF THERE WAS EVER A MOMENT in my conscious existence when I had to be committed to what I was doing, it was when the rocket I was sitting on ignited. Assuming all went well, I would ride that rocket all the way to space, accelerating from a standstill to a speed of twenty-eight thousand kilometres per hour. Only a major malfunction, such as an engine failure, a loss of cabin pressure, or an onboard fire, would force us to switch to one of the abort scenarios for which the crew had trained.

Those scenarios depended on when the problem occurred and included returning to KSC (for certain problems occurring early after launch), crossing the Atlantic and landing on a runway in Spain (or in some cases Africa), landing at Edwards Air Force Base in California after almost one complete orbit of Earth, or reaching space but at a lower altitude than planned. None of these scenarios had ever been put to the test before my flight (and never were), except in JSC's simulators, and none could be activated during the first two minutes of flight while the solid rocket boosters were burning. During that time, my crewmates and I were in the hands of the gods. Once the boosters ignited, they had to burn all their fuel and nothing could stop that process, nor could the commander and pilot take control of the vehicle. A control problem at this stage was considered catastrophic.

When the boosters separated, we were at an altitude of about forty-eight kilometres, travelling at more than five thousand kilometres per hour. We continued our ascent with the orbiter's three main engines, which are fed with liquid hydrogen and oxygen from the external tank. The good news was that we now had more control of the vehicle. Should an engine begin to malfunction, it could be shut down by interrupting its fuel supply. The bad news was that it would probably mean an abort scenario since the vehicle might not be able to reach the desired orbit with the two remaining engines. Assuming all went well, though, and all three engines maintained their normal thrust, our trip to space would take about eight and a half minutes.

Inside the vehicle, the noise was deafening for the first two minutes while the boosters were burning. Fortunately, I was wearing a helmet with a headset, allowing me to hear what was going on and, if necessary, communicate. In addition to the roar from the solid rocket boosters and the orbiter engines, I could also feel the powerful vibrations transmitted through the shuttle stack, right into my crew seat. After booster separation, it became a much quieter ride, and I breathed a sigh of relief, knowing we now had some control over the vehicle.

Later, as the shuttle continued to accelerate, I experienced a growing force on my body. That force was sustained at a level close to 3 g for the last minute before engine cutoff. Fortunately, it was experienced through my chest, so there was no risk of passing out, although I felt three times heavier than normal, which became clear when I tried to move any part of my body.

I felt nervous, of course, but it was too late to worry. Like an actor getting ready to appear on stage or a runner about to burst out of a starting block, I felt adrenaline course through my body, but once we launched, I had to concentrate on what was happening in the moment. That's what astronauts are trained to do. That's why we spend so much time preparing for all the possible scenarios. At T=0, we had crossed the Rubicon, and what mattered from that moment onwards was to keep a clear head and focus on what Mission Control was saying and what the cockpit numbers—not to mention our physical senses—were telling us. In my case, there was nothing for me to do except to listen carefully to the voice traffic. If a problem developed, I needed to understand what had happened. There would be no time for explanations.

When we reached orbital velocity—the velocity at which the vehicle can remain in space—the three main engines shut down. It's a dramatic moment called main engine cutoff, or MECO. Suddenly it was eerily quiet, and I felt myself floating loosely in my seat. It was an extraordinary feeling, partly because my body had been released from earthly forces and partly because I knew I'd made it this far, intact.

I remember looking beside me at Paul, and we both had big smiles on our faces. Nothing needed to be said. We were experiencing the same emotion—the exhilaration of knowing we were now in space. Meanwhile, Mission Control was confirming that the launch sequence had gone as planned and it was now time to transition to the on-orbit phase. Soon, I would unbuckle my seatbelt and float over to the side hatch window to experience my first view of Earth.

I thought about whether the launch had been anything like I'd expected. The answer was no. Some experiences are so unique, they can only be partially simulated and nothing can produce in you the emotional state of mind you experience when you live through the real thing.

For the record, my resting heart rate in those days was about fifty-two beats per minute. Being instrumented for this flight, my heart rate could be measured from before launch until after MECO. In the final minutes before liftoff, it was steady at 80 bpm. I was obviously excited! When the solid rocket boosters lit up, it climbed rapidly to 130 bpm and then came back down equally rapidly after MECO. I was like a human barometer of emotions.

Several things happened in quick succession once we reached orbit, beginning with the separation of the large external tank, which would re-enter the atmosphere. Engine burns were also performed with the orbital manoeuvring system to achieve a circular orbit at the desired altitude of about four hundred kilometres, and the payload bay doors were opened, which is important for managing the temperature of the orbiter.

By now, I had disconnected from my seat and floated over to the small side hatch window to get my first view of Earth. I was fully expecting to see it in the bottom half of the window, but instead it appeared in the top half because we were flying upside down! I wasn't used to such a view, so I gently rotated 180 degrees (feet towards the ceiling) in order to view it below me. Fortunately, in weightlessness everything was possible.

While every part of me knew I was in space, it wasn't until I saw Earth from that window with my own two eyes that it truly sank in. I wasn't looking at a picture. I wasn't imagining it. I was floating high above Earth, looking down at my planet, a moment seared forever in my mind. Barely twenty-five minutes after launch, we were already over Europe, having crossed the Atlantic Ocean. I could see more than 1,500 kilometres to the curved horizon. We would orbit Earth roughly once every ninety minutes (sixteen times a day), at a velocity of eight kilometres per second, and would experience a sunrise and a sunset on each orbit. Despite our velocity, we seemed to be moving in slow motion as I tried to recognize the features below.

I have often used the word *euphoria* to describe the moment I first saw Earth from space. The view that greeted me left me not only breathless but speechless. Words like *incredible, amazing,* and *extraordinary* couldn't do justice to what I was seeing or the emotions I was feeling. Words couldn't capture the power of the moment. Did the right words even exist to describe what I was feeling? Are there experiences in life that are beyond extraordinary, eluding the power of words? At the back of my mind, I knew that when I returned to Earth, I would have to describe to Canadians what I had felt, or at least try to. To this day, I struggle to describe what it's like to see Earth from space.

As we vaulted gently over our planet, I couldn't take my eyes off the view unfolding below me. I was completely mesmerized and could have remained there forever. I had begun to recognize some prominent landmarks and was eagerly anticipating the views to come when something inside me reminded me that there was work to do.

As a child, I had sometimes flown in my dreams, able to spring up from the ground and miraculously remain suspended above it as I moved around effortlessly, moments of sheer joy until I woke up and realized that I was still shackled to the ground. But now it wasn't a dream—I really was floating in mid-air. I could gently push off any surface, and that small force would move me in the opposite direction until my body made contact with another surface. My feet never needed to touch the ground.

As I pushed away from the window, I checked myself for motion sickness. I had agreed to report any symptoms I experienced. During that first hour, I felt only a mild dizziness if I turned my head too quickly. Before my flight, I had been asked whether or not I wished to medicate. If I did not, I ran the risk of being sick, preventing me from being at my best when performing my experiments. If I did, I knew it might mask symptoms I would otherwise be able to report. After consulting with the flight surgeon, I decided to medicate lightly, and only for the first two days. In addition, I would move cautiously and avoid sudden head rotations. As a result, I experienced only mild symptoms, and these disappeared after a couple of days. (For the record, I was never sick on any of my flights, a question of enduring interest to Canadian media.)

Over the next eight days, I would conduct my experiments and do my share of crew tasks, such as preparing meals and taking pictures. We were each given a Walkman (we were still in the analog era) and allowed to bring six audiocassettes. Perhaps influenced by Stanley Kubrick's *2001: A Space Odyssey* and its Richard Strauss waltzes, I chose classical music, mostly baroque—Bach, Handel, Respighi, Vivaldi, Pachelbel, Marcello—anticipating that my listening enjoyment would be even greater in space and that such beautiful music would complement the grandness and beauty of my surroundings.

Listening to Handel or Bach while floating, with my eyes closed or looking at planet Earth, was indeed a sublime experience.

I was fortunate that my flight was at an orbital inclination of fifty-seven degrees. That's the angle between the plane of our orbit and the plane of the equator. This high inclination meant we would over-fly Earth between latitudes 57° north and 57° south, while the view below us changed with each orbit because of Earth's rotation. Normally, shuttle flights were at lower orbital inclinations, meaning that a country like Canada would not be overflown. There was the added possibility of seeing the northern lights (aurora borealis) or the southern lights (aurora australis), the spectacular ovals of dancing light above each of the magnetic poles. As it turned out, the night conditions were optimal in the Southern Hemisphere. On one orbit, as we approached our most southerly latitude, I spotted the aurora australis. I was alone on the flight deck at the time and called the rest of the crew to join me for the spectacular light show. What an incredible sight!

Our daily schedule included eight hours for sleep. Because we experienced a sunrise every ninety minutes, we darkened the mid-deck and flight deck using window covers. Sleeping in space is not like sleeping on Earth and, as we all know, some people sleep better than others. If you like a pillow behind your head, you need to secure it with a strap across your forehead. There are those who also like to feel weight on them, from a blanket or duvet, but that's not possible in weightlessness. Fortunately for me, sleeping came easily, and I would tether my sleeping bag to several anchor points in the mid-deck. That worked well, although one night, just for fun, I decided to sleep untethered. This required me to stabilize myself as much as possible beforehand. You can do this reasonably well but not perfectly. You will always have some small residual movement, although it will be slow relative to your surroundings. Over time, though, you will end up somewhere else. Which is what happened to me. When I woke in the middle of the night because of a backache (a common occurrence for about half of the astronauts), my face was about nine inches from David Leestma's. I remember thinking it would be awkward if he opened his eyes at that precise moment. Fortunately, he didn't, and I regained my original position to resume my sleep.

Backaches in space occur because of changes to your body. You grow taller without the effect of gravity to compress you, as it does on Earth. I think I grew a few centimetres in weightlessness. This can result in lower back pain for some, including me. While it was mildly uncomfortable, and would wake me during the night, it was not a serious issue. I could make it go away if I deliberately compressed myself, somewhat like compressing a spring, using my hands to push on a hard surface while my feet made contact with another surface. Fortunately, it was not a problem during the day when I was constantly moving around.

Weightlessness does other things to your body. Most noticeable is the "puffy face and chicken legs" effect, whereby fluid such as blood and interstitial tissue fluid shifts from your lower body to your upper body, making your face look fuller and your legs skinnier. This can change your facial appearance and cause head congestion. Fortunately for me, I did not experience any head stuffiness and my face did not change that much since, unlike a thin person, I already had a full face.

Nor did my appetite in space diminish. In fact, I was hungrier than usual. Lucky for me, astronauts are provided with lots of food. Fresh food such as carrots, bananas, apples, and soft tacos had to be eaten in the first few days because the orbiter did not have a refrigerator. Most of the other food was dehydrated and was packaged in sealed containers to which you added water through a syringe. Food could also be warmed in a small convection oven. Some items, such as beef with gravy, came in foil pouches and did not require rehydration. There were also cookies and brownies and M&Ms and a variety of fruit juices, as well as coffee and tea. Each crewmember selected their own menu prior to the flight, with each item being colour coded with a dot to identify whose food it was. While it was nutritious, the choice was limited and most of the food was bland, although you could add hot sauce to it. (I'm glad to say it improved considerably over my three flights.) An exception (and a favourite of mine) was shrimp cocktail, mostly because of the spicy horseradish sauce which cleared your nasal passages. As you can imagine, eating required care as you extracted food from containers with a spoon and brought it to your mouth. Otherwise, it could fly all over the place.

I was somewhat surprised to hear that many astronauts lost weight on orbit. To my knowledge, I was one of the few to return weighing more than before the flight.

I exercised daily by running on the treadmill, using bungee cords to hold myself down, an activity recommended for all crew members to prevent deconditioning. I also did workouts with large elastic bands to exercise my arms and upper body. If I wasn't wearing a top, the sweat would simply pool on my skin and I would have to wipe it away with a facecloth. It actually felt good to exercise, given that we used our muscles so little most of the time.

President Reagan called our crew during our mission, a custom in the early days of the shuttle program. When the call came, the connection was intermittent, making it difficult to understand what he was saying and to whom he was speaking. We had to do some guessing, and I was concerned that he might address me and I wouldn't notice. When he did just that, I didn't realize it at first. But as the crew and I strained to hear what the president was saying, I managed to decipher the words "three strong Canadian arms," no doubt in reference to me and the Canadarm, so I jumped in and told him that I was not only proud to be part of the crew, I was also proud of the cooperation between our two countries. It was lucky I responded, as President Reagan *had* been speaking to me—otherwise I'm not sure how it would have come off to the millions of people listening, especially in Canada, if his remarks had been met by total silence on my part!

In the early days of the shuttle program, space missions generated a great deal of public interest. In addition to a call from the president, it was customary to hold a press conference for journalists in Houston and other NASA centres, where they could ask questions of the crew, all gathered in the mid-deck. These once-per-mission events lasted about twenty minutes. Although I didn't realize it at the time, my brother Philippe, who was reporting for Montreal's CKAC radio station, was the first Canadian to ask me a question. Unfortunately, I didn't recognize his voice, given the poor quality of the uplink and a signal dropout at the beginning when he was identifying himself. Consequently, I didn't acknowledge that it was great to be talking to my own brother. His question, slightly garbled and which he first asked in French and then again in English, was: "Marc, in your own words, what were your impressions of takeoff?" (After the flight, Philippe told me that CBC and CTV had argued about who should be the first broadcaster to ask me a question and, given that neither could agree, they offered him the opportunity to go first.)

Because of the large Canadian media presence, I was asked six questions, more than my share considering the crew size. Given the limited time available, I tried to answer them without monopolizing the microphone, conscious that I was with six other people, with lots of "mission firsts" (largest crew ever, first American woman to do a spacewalk, and first flight with two women) and only twenty minutes at our disposal. Unbeknownst to me, I was dubbed "the Right Stiff" in the *Toronto Star*, partly because I hadn't acknowledged talking to my own brother and because the Canadian media wanted to hear much more from me about what it was like to be in space. I laughed when I heard this and resolved to be more expansive in my postflight tour, when there would be fewer time constraints.

While in orbit and in the years since, I have often thought about how your perspective in space differs so dramatically from when you're on

Earth. Considering how few people are able to weigh in on the subject, I paid particular attention to my feelings about it.

On Earth, your visual horizon is the small circle around you, extending out to perhaps ten or fifteen kilometres. Even though that circle moves with you as you travel, your physical world remains small. But when you're orbiting Earth once every ninety minutes, your visual and mental horizons extend dramatically—the entire planet becomes your world. This makes it hard not to think of larger, more global issues. At least that was my experience. In space, I thought a lot about Earth's atmosphere, which surrounds our planet, and its oceans, which cover 70 per cent of it, all of which we share, knowing both are under threat. I thought about how we have divided our planet into about two hundred countries, many of which do not get along with each other. I wondered where humanity was headed. My "local" priorities became much less important, at least for a while. My perspective while in space shifted in a big way, untethered from the realities of my day-to-day life, and broadened by seeing my planet from such an amazing vantage point.

This is what drew me to a window whenever I had a spare moment. I wanted to experience it as much as possible. While our planet is breathtakingly beautiful, I remember observing the wafer-thin atmosphere that made life possible, and thinking that we can't afford to damage it any further. I remember seeing a million square kilometres of smoke dispersing above the Amazon forest as it was being deliberately burned to convert it to farmland. I remember seeing red earth being washed "like blood" (as someone described it) into the Indian Ocean from Madagascar's Betsiboka River because of deforestation and the resulting soil erosion. I remember seeing great palls of yellow-brown air over California, over the Mediterranean, and over China, to name a few locations. There is no question in my mind that we are damaging our planet, sometimes unwittingly, sometimes deliberately. And yet, planet Earth is the only home we have, a home we must all share because there is nowhere else for us to go. It is the cradle of humanity, surrounded by the immense void of space. Its fate rests solely with us. From the vantage point of space, you can't help but think and feel this in your bones.

When we are young, we have an imagination that allows us to think that almost anything is possible. Our thinking is not bound by the laws of physics. As we grow up, we begin to lose that kind of almost pure imagination as our parents and teachers tell us about the real world and we learn that some things are just not possible. Some of us hold on to our imaginations longer than others, but we all eventually realize that on Earth we cannot do certain things, such as float across a room suspended in mid-air. But guess what! In space you can. It is a place where once again you become a child and redis-cover the imagination you lost by growing up. This explains why seri-ous adults laugh like children when they first arrive in space.

Space is experienced on four levels: physical, intellectual, emo-tional, and, for some, spiritual. The first three are obvious; the last, less so. I had echoed the thoughts of an American astronaut who once said: "When I sat in the shuttle waiting for launch, I hoped there was a God. When I got to space, I knew there was one." As someone who has struggled with his faith, I could relate to the first part of this, although I still struggle with God's actual existence.

Seeing our planet as I did certainly reawakened in me that age-old question of how it all came to be, but this time in a manner that was more pressing and compelling than ever before. A part of me wants to believe that some divine being made it all happen. There is no ques-tion that being in space intensified my curiosity about the origin of the universe and whether we'll ever get the answers we are seeking.

After eight days and 133 Earth orbits, it was time to come home. It had gone by so quickly, and I wondered whether I had taken the time to savour it. While my work had to come first, it was also important for me to absorb the singular experience of being in space. I did not want to be asked on my return what it was like and have to answer, "I don't know, I was too busy to notice."

As I've already said, I was incredibly lucky to be on a high-orbital-inclination mission focused largely on Earth observation, with *Challenger*'s overhead windows facing Earth during most of the flight. Such missions were rare, and allowed our crew to enjoy an unparalleled view of our planet. I had the pleasure of seeing different parts of my country on many orbits, allowing me to appreciate not only its sheer size but also the natural beauty of its mountains, lakes, rivers, forests, plains, and coastlines. It made me realize how fortunate I was to live in such a country.

For me, orbiter re-entry illustrates better than anything else the technological marvel that was the space shuttle. Think of it: a space vehicle that decelerates from Mach 25 and glides back to Earth to land on a runway. While perhaps less visually dramatic than the launch, re-entry is, to my mind, the more impressive phase.

Unlike at launch, I did not experience any butterflies before re-entry, perhaps because the process does not involve the release of tremendous amounts of energy. In fact it's the opposite—it's about bleeding off energy. This requires that the thermal tiles, blankets, and special carbon coatings surrounding the aluminum body of the orbiter remain intact. This had been the case on previous flights, giving me peace of mind. Tragically, the loss of the orbiter *Columbia* during re-entry almost twenty years later would illustrate what happens when that thermal protection is compromised.

At the end of our fantastic voyage, I was feeling serene. This serenity came from getting done what I had set out to do. I felt I had lived up to my own expectations and my crew's, as well as those of my country.

In preparation for re-entry, we returned to our seats and buckled up, and after rotating into a tail-forward position for a deorbit burn lasting a few minutes to slow us down, we rotated back to nose-forward and began our descent. We were now falling out of the sky, going too slowly to remain in space. Unlike at launch, there was no noise and vibration. Gradually, we began to plough into Earth's thickening atmosphere. Re-entry is all about energy management and the need to bleed off speed in a precise manner to arrive at the threshold of the runway at Kennedy Space Center with a touchdown speed of roughly 200 knots (about 370 kilometres per hour), plus or minus a few knots depending on the weight of the vehicle. Since the orbiter is a glider, the pilot had only one chance to get it right. We couldn't go around for a second try.

During re-entry, which takes about an hour, I gradually began experiencing the effect of gravity—which goes from 0 g to a maximum of about 1.6 g—and it felt strange after being weightless for eight days. My body had quickly adapted to floating, and now I had to reacquaint myself with heaviness. I held a pencil in front of me and watched it slowly "sag" towards the floor. For a brief time I had forgotten what gravity was like, and now it was returning in full force, and it surprised me.

Challenger landed at KSC at 12:26 p.m. on October 13. Shortly after "wheels stopped" and a few orbiter "safing" procedures, a mobile staircase was wheeled up to the vehicle, the side hatch door was opened, and a flight surgeon came onboard to see how everyone was doing. We then exited the vehicle and descended the stairs, most of us holding on to the rail until we reached the ground, given that we hadn't used our legs for some time.

As was customary, we walked around the vehicle to see how it looked. Apart from some insulation missing from the right orbital manoeuvring system pod (reported on day 1 of the flight), there were some damaged thermal tiles, including a missing one on the underside of the left wing. Otherwise, *Challenger* looked good.

It had been my home for eight days. I could not help but feel a special attachment to it, perhaps in the same way as I had felt a continuing connection after serving in HMCS *Algonquin*. An extraordinary part of my life had unfolded inside the shuttle, an experience I would never forget.

The sun was shining, and while it felt a little strange to be back on Earth, I was looking forward to seeing my family, who were waiting for me in Houston. I also wondered whether I would have a chance to fly in space again. Once you've been there, you want to return. How could you not? Regardless of what the future held, I was conscious that I had been blessed beyond my wildest dreams.

Following our walkaround, we returned to crew quarters to shower and undergo a medical checkup. Showering after eight days of using nothing more than a washcloth was an exquisite experience. Shampooing my hair was even more satisfying. We then flew to Houston, where I was reunited with my family. I felt incredible joy to be with them again. I had not seen Simone and Yves in more than two months. All my crewmates were there with their own families and everyone was thrilled that the mission had gone so well. After a few short speeches, we gradually dispersed as everyone headed home. None of the crew were allowed to drive so soon after landing, so my brother Philippe did the honours, and because none of us had eaten, he picked up some Kentucky Fried Chicken. "From KSC to KFC" was my brother's memorable line that day.

Once back at our apartment, I asked my family how it had felt to watch the launch. Jacqueline described everyone's excitement, but confessed to some apprehension watching the liftoff and seeing us rise into the sky until we disappeared. As an adult, she understood the risk. Not so for Yves and Simone, who thought it was pretty cool but were more excited that they had gone to Disney World in Orlando right after the launch. Of course, I was happy my kids got to be there, but I was also glad they were too young to worry about me.

Obviously, I wanted to share every detail of my experience with my family, and they did their best to listen to my enthusiastic monologue. I was still on a high as I described each moment and how it felt, particularly how my perspective had shifted to the bigger challenges faced by humanity. I could have talked for hours, but I noticed everyone was tired after a long day, and so I suggested we should all go to bed. Jacqueline, looking grateful, turned to me and said: "Honey, that was really interesting, but before you come to bed, would you mind putting out the garbage?"

FIVE

ALTHOUGH MY SPACEFLIGHT HAD ONLY LASTED EIGHT DAYS, I knew every moment of it would stay with me for the rest of my life, never fading into the recesses of my memory. It was just too intense to ever be forgotten.

With my mission behind me, I could now look forward to spending more time with my family, even though the respite would be brief. Ahead of me was a speaking tour across Canada that would last until the following summer. But before that, I needed to participate in some postflight debriefings at JSC, and Jacqueline and I would also take a short holiday. We both needed a break.

It was customary for the Astronaut Office to conduct a post-mortem after every flight. Some debriefings included the payload specialists, with a focus on the mission itself and the training we received. This was my opportunity to provide feedback, and I gave it. In my opinion, the mission had gone well and the training had been excellent, but not fully integrating payload specialists with the rest of the crew before the flight was something that should be corrected. My recommendation was accepted, although there may have been a few astronauts who were not pleased, believing it was still unnecessary for NASA to fly payload specialists.

Following the debriefings, I was free to leave. After saying our thank-yous and goodbyes, Jacqueline and I flew to the Bahamas to decompress and to rest before the busy months ahead. Days before,

I had enjoyed the stunning view of the Bahamas from space. Now I would get to enjoy it close up.

More than anything, the break allowed me to reflect on my spaceflight, an experience I was still making sense of. Being a scuba diver, I dove almost every day and thought about the similarities between floating in space and looking down at Earth, and swimming underwater, neutrally buoyant, observing the sea life. It was a wonderful way to transition back to Earth.

Processing my spaceflight experience was partly about finding the right words to express what I felt at the time. Because I knew I would be asked, I wanted to answer the question of whether the experience had changed me. It's a question I am still asked today. This is what took shape in my mind: Yes, spaceflight does change you but not in a dramatic or strange manner, at least not in my case. For me, the change occurred in two ways: first, in how I viewed planet Earth and the challenges facing it, and second, in how I viewed my own life. I now saw the future of our planet from a more global perspective. Quite literally, the issues that now mattered most to me were planetary ones: climate change, loss of biodiversity, and the tensions that threaten world peace and security.

The second change was about how I viewed my own future. Not that I had ever felt insecure, but life does seem less daunting after completing a space mission. You have had to prepare for something that carries the risk of death and come to terms with that. You have also accomplished what you set out to do. The resulting peace of mind is an unexpected gift. I believe it has enabled me to deal more calmly and patiently with some of the challenges and difficulties I have faced since then.

While Jacqueline and I were relaxing, the National Research Council was busy preparing my postflight tour. It would begin the evening I returned to Canada, with a public presentation in Hull (now Gatineau), across the river from Ottawa. A Challenger aircraft would pick us up in Eleuthera and fly us to Ottawa after a one-night stopover in Washington, D.C., where I would review the NASA mission film I would be presenting to the public. I needed to see the film beforehand, to create the bilingual narrative I would deliver while showing it. The Challenger also brought along my parents and children because the media would be waiting for all of us on the tarmac when we landed at Ottawa.

Returning to Canada was an emotional moment for me. I was overwhelmed by the warmth and enthusiasm of those who greeted me and my family. Canadians were eager to hear my story. That evening, I stood before a general audience and described my mission for the first time. The enthusiasm I sensed in the audience, as they listened and later asked questions, energized me and made me realize just how much Canadians were fascinated by human spaceflight. I was also deeply touched by how many told me they were proud of what I had done.

My presentation went well except for one interruption. Because we were in the National Capital Region, I had been asked to speak in an equal mix of French and English. This was not acceptable to one audience member, who shouted from his seat that I was in Quebec and should speak in French only. Not surprisingly, this caused some of the English audience to grumble. What can I say? I had scrupulously adhered to the fifty-fifty formula, knowing half my audience was anglophone and had travelled across the Ottawa River to hear me speak, and someone in the audience wasn't happy about that. It's hard to please everyone, especially when it comes to Canadian language issues!

Soon after my return, I received a letter from a publisher proposing I write a book about my flight. They even offered a ghostwriter. It caught me by surprise and I discussed it with Jacqueline. In the end, I said no. With so much ahead of me, now was not the time to write my story, though I might consider it when substantially more of my life was behind me.

Although I was not ready to become an author, I was certainly ready to tell Canadians about my flight. I would do this pretty much non-stop for eight months, except for a short break at Christmas. I viewed it as a serious obligation. I had been given the opportunity to fly in space and now it was time to share that experience.

Although I didn't know it at the time, I was destined to be the only flown Canadian for the next seven years, given the tragic loss of the *Challenger* in 1986. This meant giving hundreds of presentations to the public, sometimes in classrooms, sometimes in large public auditoriums, sometimes in church basements, where I would do everything, including threading my own 16 mm film on the projector, loading a 35 mm slide carousel, and operating both as I spoke to my audience. It partly explains why, to this day, I retain such a clear memory of my first flight.

Before the start of my tour, my family was invited by Prime Minister Mulroney and his wife, Mila, to lunch with them on Parliament Hill. For the record, the conversation was about space, not politics. Not surprisingly, the prime minister and his wife asked the same kinds of questions as the general public, such as what the launch feels like and the experience of seeing Earth from space. They were also interested in hearing what Jacqueline and the children thought of it all. They also asked about the future of the Canadian astronaut program, including opportunities for Canadians. Naturally, I expressed the hope that many more would get to fly.

Personally, I have always liked Brian Mulroney, and even though I would later become a Liberal MP, my values are not far removed from those of Progressive Conservatives like him. I think he made a significant contribution to our country.

Around the same time, Bob Crippen, my mission commander, was invited to Canada and we both attended Question Period in the House of Commons, viewing it from above, in the gallery. It was my first visit to the House, and the exchanges that day were particularly acrimonious. As a spectator, I was embarrassed by the theatrics of certain MPs who didn't stop heckling. I wondered whether Bob felt the same way. Did the unruliness on display below shock him, casting doubts on the widely held American view that Canadians were a polite and courteous people? I remember turning to him and apologizing for all this childishness. Embarrassed or not, without skipping a beat, he said: "Don't worry, Marc. It's the same in our Congress!"

At the time, of course, I didn't know I would take my own seat in the House, twenty-four years later. Notwithstanding the heckling and shouting, I was impressed with the precise clockwork of Question Period as roughly thirty-five MPs stood and were each given thirty-five seconds to ask a question. As a spectator, it was a challenge for me to follow it. Did it make me want to consider running? Although the thought had crossed my mind more than once, I felt that going into politics was something I might want to attempt later in life, when my astronaut career was behind me and I could bring more experience to the table.

It was now time to go on tour. It began with protocol visits to each province, where I met the lieutenant-governor and presented a photo montage that included a small provincial flag that had flown aboard *Challenger*. I also met the premiers. And of course I made presentations to the general public and to schoolchildren. This is what I enjoyed the most, seeing faces light up in wonder as I described the experience of floating and of seeing Earth from space.

I had to adjust my presentations depending on the age of the children. Obviously, I couldn't talk the same way to a kindergarten class and a high school class. When both were in the auditorium at the same time, my presentation skills were put to the test.

Over time, I refined my pitch, adding more details, injecting humour, and fine-tuning my answers to questions like: "How do you go to the bathroom in space?" which I never failed to get, mostly from children, no doubt prodded to ask it by their equally curious parents. I would point out the obvious—that because of weightlessness, a flush toilet with water would not work. Everyone seemed to understand that. Many would start to giggle when I explained the process. If you had to go to the bathroom, you sat on a special toilet, secured yourself to it so that you wouldn't float away, and opened the normally closed receptacle on which you were sitting. You then did your business and relied on suction to separate your waste from you, although you double-checked when it was time to wipe your bottom (the suction was not very strong). You then resealed the toilet to make sure nothing floated out. For urinating, you used a hose to which you attached a funnel device (we each had our own). It also used suction to draw the urine into a receptacle. It was a cumbersome process although I have to say it got the job done (pun intended).

A constant question from young people was: "What should I do if I want to become an astronaut?" The short version of my answer touched on three points: the importance of an education in science, engineering, or medicine; staying healthy by exercising and adopting good nutrition; and, finally, developing public speaking skills, because astronauts had to be able to communicate clearly. What I didn't say is that they should always keep an eye on the Help Wanted section of their local newspaper!

My tour of the capitals began with my birthplace, Quebec City. This stop included attending an event at city hall with Mayor Jean Pelletier, having lunch with provincial politicians at the National Assembly, meeting Premier René Lévesque (a man I liked, despite our different views about the future of Quebec), and riding a Zamboni at half-time during a Nordiques game!

During one visit to Toronto, I was joined again by Bob Crippen for a presentation at Ontario Place, where we narrated our NASA mission film before a general audience. Bob and I alternated as we described the highlights of the flight. Afterwards, one audience member stood up and said that while he had enjoyed listening to us describe what happened on the flight, he was also interested in the emotions we felt as we lived that experience: "You told us what you did during your mission, and that's good, but you didn't tell us how you felt. Gentlemen, where are your souls?"

With that, I realized that many people were seeking to live the experience of spaceflight vicariously through me and other astronauts. From then on, I made a point of sharing my emotions more openly in my presentations, describing what I felt when I first saw Earth from space, how I dealt with the risks of spaceflight, and what I learned about myself and others while I prepared for my flight and while I was in orbit.

I also continued to do media. One of my first radio appearances was on CBC's Sunday afternoon program *Cross-Country Checkup*. I had been warned that, although most of the questions would be friendly, there might also be some difficult ones. I remember two in particular: "With all the poverty in the world, why are we spending millions of dollars going to space?" And, "Why are we cooperating in space with the U.S. when they want to fly cruise missiles over Canada's North?" (a controversial issue at the time).

I realized that while most Canadians just wanted to talk about space, there was no way to avoid political questions. Responding to them came with being a public figure and turned out to be valuable training for my subsequent career in politics.

On the subject of why Canada should spend money on space, it was a valid question that I still get asked today. The first part of my answer was straightforward as I pointed to the value of satellites that allow us to communicate with each other over vast distances, that allow us to locate our precise position anywhere on the planet's surface, that allow us to monitor and predict the weather, and that allow us to observe our planet and measure some of the effects of climate change. When Canadians are made aware of this, most react positively, recognizing the benefits.

The second part of my answer was more complicated because it dealt with sending humans to space, something not everyone agrees is worthwhile, especially when they argue that it would be cheaper and safer to send robots.

So, let's look at robots. Everybody agrees we should send robots to places where, for safety reasons, humans can never go. In some cases, it also makes sense to send robots on precursor missions before we send humans—for example, in the case of Mars exploration, where robots can conduct scientific experiments, helping us to learn more about the planet. Finally, with advances in artificial intelligence, it may also make sense to use robots when we feel confident they have the ability to act independently, particularly when they're too far from Earth to be monitored or controlled in real time by a human operator in Mission Control.

That said, robots can't talk to us or inspire us the way humans can (at least not yet), and people want to be inspired by their fellow humans. Robots can't experience and convey the complex range of emotions that humans feel. R2-D2 and C-3PO are cute, but they're still machines. When I spoke about my spaceflight, people wanted to be there with me, experiencing what I was experiencing. In the past forty years, I have spoken to countless children, so excited that they were tripping over their words as they tried to get their questions out. I have often seen that same excitement in adults. When we are inspired, it stirs our imagination, which is key for humanity.

When my cross-country tour slowed down, life returned to normal, and Jacqueline and I were able to focus on buying a house. We knew we would be living in the Ottawa area for the foreseeable future and preferred to own rather than rent. We were fortunate to find a place three blocks from where we were living. The children could continue at their school, and all our neighbourhood friends would still be nearby.

When I was selected as an astronaut, I was allowed to remain in the navy, even though I would spend all of 1984 preparing for my spaceflight. Early on, I met with Gen. John de Chastelain, at his request, to discuss my future. I told him that I would be honoured to continue wearing the uniform, but that I now wanted to work full-time with the Canadian astronaut program. He accepted this on the condition that I would continue to promote the Armed Forces as circumstances permitted, something I was happy to do. I would proudly wear the uniform and hold my commission as an officer until 1989, when I retired as a naval captain.

After my postflight tour, I was able to rejoin my colleagues and focus on the next Canadians to fly. NASA had originally promised us two flights, but we were now looking at three. This was in part because Canada had signalled its interest in participating in the International Space Station program being promoted by President Reagan. I was pleased that two more Canadians would get to fly and that the door might be opened to even more flights. But as Canada prepared for these flights, tragedy struck.

I had gone down to the Johnson Space Center in late January 1986 to discuss one of the flight opportunities for which we were preparing. Seven weeks before, Canada had named Steve MacLean to fly on a mission in which he would perform a suite of Canadian experiments known as CANEX-2, with Bjarni Tryggvason as his backup. NASA was exploring several options for his flight, and I was meeting with JSC payload manager Charles Chassay to discuss the details.

As the countdown proceeded to *Challenger*'s tenth launch, we all paused to watch it on television. One of the crew of mission STS-51L was Payload Specialist Christa McAuliffe, a schoolteacher. Her presence on the flight had generated enormous interest, in part because of her engaging personality and because she was seen as uniquely qualified to educate the public about space and what makes it so different from Earth. She had captured the hearts of Americans.

As the shuttle cleared the launch pad, I felt the same exhilaration we all feel watching it rise into the morning sky, marvelling at the human ingenuity that had created such a machine. Like everyone else in the room, I couldn't take my eyes off it as it climbed higher and higher for seventy-three seconds. And then, in a fraction of a second, it exploded and came apart.

Suddenly, the world no longer made sense. Like everyone else, I continued to stare at the screen, not wanting to believe what I had seen. Because I had flown and knew precisely what was supposed to happen every second of the flight, I knew there had been no option for the crew to escape. Despite that knowledge, I kept staring at the screen, praying for a miracle and listening to Mission Control, hoping they would tell us something other than what we already knew. But even Mission Control had gone silent as we all watched the cataclysm unfolding before us. There were gasps in the room, and some people were crying, but there was also denial. This could not possibly have happened. But it had.

There were seven astronauts onboard, and they were each, in their own way, exceptional. One of them, Judith Resnik, who had spent time in Canada developing the operator procedures for the Canadarm, had reached out to me during my training to make me feel welcomed. The world had lost seven wonderful people.

In due course, the full details of what had happened became public. On the morning of the launch, the temperature at KSC was colder than usual. The two O-rings (one primary and one secondary) that provided seals at the joints between the mated segments of the solid rocket boosters had become harder at the lower temperature, losing their sealing properties and allowing hot gases to shoot out sideways between two segments after booster ignition. Acting like a blowtorch, these hot gases destroyed the lower strut attachment securing one of the boosters to the large external fuel tank and burned into the tank itself. This led to the loss of structural integrity of the shuttle stack and the massive fuel tank explosion that followed.

There was much speculation about what the crew experienced after the explosion and the separation of the crew module from the orbiter. It is probable that some survived the initial separation, although surviving the subsequent ocean impact was never a possibility.

I cannot fathom what it must have been like for their families and friends who were watching the liftoff, either in person or on television.

To experience such a catastrophic event was hard for everyone; in their case, the tragedy must have left scars that would never fully heal. It was a moment when we witnessed the unforgiving nature of machines if we operate them outside their limits. And it was also a moment to reflect on how precious human life is, and how quickly it can be extinguished.

While I have already said that I had to make my own assessment of the risks involved in spaceflight, and I did so before my own flight, I never dwelt on the specific details of what could happen to me and my fellow crew members in various catastrophic scenarios. That would not have been a productive exercise. Nevertheless, the *Challenger* disaster was a tragic and graphic reminder that such scenarios were possible.

As recovery efforts began, NASA announced a memorial service at JSC on January 31. President Reagan flew to Houston to address the nation at a sombre but moving ceremony. The families and friends of the seven astronauts gathered on the grounds of JSC, along with international delegations from other spacefaring nations. President Reagan gave an inspirational speech and quoted from the famous poem "High Flight" by John Gillespie Magee, Jr. ("Oh! I have slipped the surly bonds of Earth . . ."). He spoke of the indomitable American spirit, compelling it to take on challenges, a quest that must continue.

As part of the Canadian delegation I would join Pat Carney, minister of energy, mines, and resources, and Frank Oberle, minister of state for science and technology. Even though it was a sad occasion, I was glad to be there. I felt a strong kinship with my fellow astronauts and wanted to mourn with them.

I flew down with the ministers the night before the memorial, and we spoke of the *Challenger* disaster and the investigation that would follow, as well as about the future of the Canadian space program and of difficult government decisions ahead. At the time, Canada was considering joining the International Space Station program, but also thinking of investing in the design and construction of an innovative

Earth observation satellite called RADARSAT-1. Both were expensive programs. The question was whether we could afford both. Minister Carney was responsible for the RADARSAT program, and Minister of State Oberle, for the ISS program.

RADARSAT was a new technology that enabled Earth observation using a special radar, called a synthetic aperture radar, that could see Earth's surface, day or night, regardless of cloud cover. It would prove invaluable for a large country like Canada wanting to monitor its land, lakes, rivers, and territorial waters. The International Space Station, on the other hand, was an ambitious program to build a permanently inhabited station to orbit Earth. Potential partners included the U.S., Russia, Japan, Canada, and the countries of the European Space Agency. If Canada participated, it would open the door for us to build the station's robotics, allow our scientists to fly their experiments, and create more opportunities for Canadians to fly.

I should add that any hesitation on whether to finance our participation in the International Space Station had nothing to do with the tragic loss of the *Challenger* crew. It was strictly a question of whether Canada was willing to spend the money required to participate.

As we flew to Houston, the ministers asked me point-blank whether I thought one program was more important than the other, given that Canada might not finance both. (A third program for communication satellites called MSAT (Mobile Satellite) was also on the table but not discussed).

While I recognized the importance of having a leading-edge Earth observation satellite, I was also hoping that Canada's astronaut program would have a future through our participation in the ISS program. I knew this was a rare opportunity for me to speak my mind to two important decision-makers and I chose to argue for both. I won't claim that my pitch influenced them, but happily, Canada made the farsighted decision to go ahead with both programs.

Having flown in *Challenger* and then witnessed its destruction, I was asked by friends and media whether I'd ever want to fly again. My answer was an unequivocal yes, and I believe the great majority of astronauts responded the same way, especially those who had not yet flown. The truth is that astronauts know the risks of spaceflight but are also drawn to the profession because of their strong desire to be on the edge of what is possible. As for my family, none of them ever said to me that I should not fly again. The subject was never raised.

Needless to say, NASA's space shuttle program ground to a halt after the *Challenger* disaster. Some were questioning whether the shuttle program should continue and, if it did, how long it would take before flights resumed. In the immediate aftermath, no one knew. I certainly didn't, but I assumed NASA would do everything possible to keep the program going.

Its first step was to announce the creation of the Rogers Commission to investigate what had happened. Its chair, William Rogers, had been secretary of state under Richard Nixon and attorney general under Dwight Eisenhower. Sally Ride, with whom I had flown, would be named to the commission. The commission's findings would lead to a number of important changes to the shuttle program, including revisions to flight rules and procedures, as well as modifications to hardware, beginning with the solid rocket boosters, the primary cause of the disaster.

As the commission began its work, it became clear that there would be a lengthy hiatus in NASA's human spaceflight program. That also meant a delay in our own plans to get the second and third Canadians into space. For the time being, my back-and-forth visits to JSC were on hold. For how long, I didn't know, although I certainly underestimated how long it would take before the next Canadian flight. The space shuttle would not fly for two and a half years, and there would be several more years of delay because of backlogged flights. As a result, Canadian astronauts would not fly again until six years later, in 1992, when both Roberta Bondar and Steve MacLean would fly on separate missions.

Because we now had more time on our hands, the Canadian astronaut office decided to add skydiving to its training curriculum. This was because NASA was incorporating a crew bailout capability into the orbiter as part of its safety improvements following the *Challenger* disaster. This would allow crew members to slide down a pole that was extended through the side hatch opening and parachute to the ground or into the sea while the orbiter was still in flight and unable to land safely on a runway. Experiencing free-falling was considered worthwhile training. I would end up doing about thirty jumps from an airfield near Ottawa.

I recognize that skydiving is not for everyone. Preparing for a jump is serious business, beginning with a properly packed parachute. You must also have the presence of mind to deploy your chute at the right altitude, and if it fails to open properly, open your reserve chute, and then decide how and where to land without breaking any bones. Based on this, you may ask why anyone would voluntarily jump out of an airplane. Beyond its practical purposes, it's all about the indescribable sense of freedom you experience while free-falling.

Because Canada wanted to encourage scientists to explore the potential of doing research in microgravity on the space shuttle or the future International Space Station, we also began organizing and participating in flights on NASA's KC-135, so that scientists could perform experiments or test hardware during the brief periods of weightlessness of each parabola. These flight opportunities became so popular that even high school students participated.

During this time, I began to deal with a serious challenge related to Jacqueline's health. It began in 1985, when she was diagnosed as having bipolar disorder. Her behaviour had changed noticeably in the span of a few months, and I worried that the depressive episodes she had experienced after her brother's death in 1983 were returning. She had been remarkably well from the time we moved to Ottawa, until almost the end of my postflight tour, when I began to notice changes. I had reduced my travel to a minimum as we sought help, which included psychiatric care and treatment in hospital. Meanwhile, I was trying to ensure as normal a life as possible for Yves and Simone, particularly when their mother was away. Fortunately, we lived in a friendly neighbourhood and the children's school was literally at the end of the street. I gave them each a house key so they could let themselves in after school before I got home. In addition, my mother and some neighbours kept a watchful eye.

I had pretty much become a single parent, taking care of ten-year-old twins. Unfortunately, as time went on, Jacqueline's condition worsened, and she spent more and more time in hospital. At home, she was unable to regain any interest in the normal activities of life. She continued to be severely depressed, and nothing seemed to help. It was heartbreaking to see this happening to the woman I had loved for fourteen years. She had not chosen this for herself. Yves and Simone were also wondering what was happening to their mother.

Over time, this put a strain on our marriage and, eventually, led to our decision to separate, a sad outcome to a marriage that had given us two beautiful children and so many happy memories. I wished it could have been otherwise and I accept my share of responsibility for our mutual decision to live separately. We agreed that the children would live with each of us on an equal basis. However, because Jacqueline did not feel well enough to take them, Yves and Simone continued living with me.

Before our separation, Jacqueline had become friends with a patient she had met in hospital, and in 1987 they ended up sharing an apartment nearby. Unfortunately, he was not well.

The police came to my door the morning of Sunday, April 26, to notify me that Jacqueline and her roommate had taken their own lives. It was the worst day of my life. I couldn't believe she was gone. I needed to tell Yves and Simone, but how do you explain this to eleven-year-olds? They were both at choir practice when I got the news. I tried to compose myself and think of what I would say to them when they came home. How do you tell your children that their mother is dead, let alone that she has taken her own life? I realized the only possible approach was the simple truth. When they returned, we all sat in the living room, and I told them what had happened. We cried and we hugged. Although they had seen Jacqueline's condition deteriorate over an extended period and had asked me many times about her situation, children still assume their parents will be around forever. I could not imagine how hard it was for them to hear that they had lost their mother.

For almost two years I had felt powerless, unable to do anything that might help pull Jacqueline out of her depression. I was in touch regularly with her psychiatrist and had been kept abreast of her condition and the treatment options. These sometimes worked for a while and allowed her to re-emerge as the person I had known, but only for brief periods. Because few people wanted to talk openly about mental health in those days, I felt isolated and could only turn to my parents to share my concerns.

In the thirty-six years since Jacqueline's death, I have read a great deal about bipolar disorder and the progress that has been made in treating it, allowing people to lead fulfilling lives. It has strongly shaped my views on the importance of focusing more attention on mental health and of removing the associated stigma.

I was now a single parent and had to make it work. Fortunately, Yves and Simone were incredibly resilient. Without realizing it, they were helping me. We were supporting each other. We sat together at the dinner table every night and talked about their day at school and their friends. They never complained about the boring meals I prepared. In fact, they joked about them. In time, they became more assertive and even ganged up against me buying their clothes. Secretly, I was glad they did. I wasn't particularly knowledgeable in that department. I agreed to give them each a clothing allowance, provided they occasionally bought underwear and socks. Thank God for my mother nearby, who would also lend a hand. Gradually, we moved on with

A MOST EXTRAORDINARY RIDE

our lives and got used to being a family of three, although I know Yves and Simone dearly missed not having a mother.

Being a single parent changed me. I focused less on myself and more on my two children than might otherwise have been the case, and I cherish that period in my life. We did a lot of things together and became very close, even though I knew there was a void I couldn't fill. I became a better listener when they came to me, and when I sensed they might be reluctant to bring something up, I tried to draw them out. I was sensitive to their moods and was there quickly when they were experiencing the difficulties that all young people face, primarily in their friendships at school. Those who have been single parents will understand me when I say that my priorities in life had changed because I could no longer share parenting with someone else. It fell squarely on my shoulders. I believe it made me stronger and more empathetic.

That summer, we went on a camping trip to Algonquin Park. We had been there before and had enjoyed it. This time, the three of us set out to walk many of the nature trails in the park to learn about its animals, birds, plants, trees, and fungi. Some trails were short and some were much longer. I was particularly impressed when Yves and Simone trekked nineteen kilometres, part of it through the rain, on the park's longest scenic trail. We also swam, made campfires, picked blueberries for our pancakes, visited a lumberjack museum, and listened to a lecture about wolves. It was the first time I was with my kids all day, every day, for ten days as a single parent. I recommend it to other fathers.

Later in the year, I received a call from Dick Dodds, director of education for the East York School Board, informing me of a proposal to rename one of their schools after me and seeking my permission. Overlea Secondary School would be renamed Marc Garneau Collegiate Institute. I was both surprised and honoured. I accepted with pleasure and was invited to speak at the official renaming ceremony. It was a proud moment for me which Simone and Yves were there to

97

share, and the beginning of an enduring friendship with many of the teachers at the school. I made a point of telling the students that I was a bit nervous about having a school named after me since it meant I had to be a good role model for the rest of my life, or else the school would quickly change its name! Talk about a lot of pressure.

Over the years, I have visited the school many times and interacted with the students—always an enjoyable and enriching experience. On one occasion, a student asked me if I was really Marc Garneau. "I thought you were dead because this school is named after you," he told me. The kid had a point!

In 1993, I was honoured to have a second school named after me, this time a French school in Trenton, Ontario: École Secondaire Publique Marc Garneau, which belongs to the Conseil des Écoles Publiques de l'Est de l'Ontario. Being a strong supporter of bilingualism, I was truly proud knowing that I now had both an English and a French school named after me. That pride has only grown over time, now surpassing three decades.

After a thirty-two-month pause, *Discovery* took to the skies in September 1988, signalling the resumption of NASA's space shuttle program. Excitement for human spaceflight had been building again, although it was now tempered by the sobering reality of the risk it carried.

Meanwhile, in Canada, we continued preparing for the next two Canadian flights. The first would focus on life sciences and the changes that occur in the human body in weightlessness. Roberta Bondar would be named in early 1990 to undertake this mission, with Ken Money as her backup. The other flight would focus on space technology and the Canadian Space Vision System. Steve MacLean had already been named to this mission, with Bjarni Tryggvason as his backup. Unquestionably, these flights were far more ambitious than mine had been, given the longer time to prepare. Ultimately, they would both occur in 1992, a banner year for Canada in space, with

Roberta and Steve becoming the second and third Canadians to fly. Their flights would generate a great deal of interest. Best of all, two more of my fellow astronauts would be able to share their spaceflight experience with Canadians.

In the summer of 1989, I met Pamela Soame, the daughter of long-time friends of my parents who were members of the same tennis club. She was also a member and had been told by her parents that I occasionally played there with my own parents and that she should call me for a game, which she did. I first joined her in a doubles game and then met her again to play singles. (She won.) It had been two years since Jacqueline's death, and I was beginning to come out of my shell. Pam was a registered nurse in the Canadian Armed Forces and I liked her the moment I met her. She was full of life and very funny.

Yves and Simone clearly enjoyed Pam's company and loved her sense of humour. They often suggested I invite her over for dinner, a clear sign that they liked her. The conversation was always easy, and it was wonderful to see my children looking happy and laughing. I will be the first to say that Pam filled a void in their lives that I could not possibly fill. It was the beginning of a promising relationship. She lived nearby and I would see her regularly. While neither of us was in a hurry, our relationship grew over time. We went out to movies and restaurants and we enjoyed jogging, playing tennis, and cross-country skiing, clearly bringing out the competitive streak in each other.

One time when we were cross-country skiing in Gatineau Park, that competitive streak got the better of me. We were returning from an outing and skiing back to the parking lot. We were on a winding, descending path, and Pam, an experienced cross-country skier, was leading. As we descended, she got farther and farther ahead of me, and I decided to throw caution to the wind and ski faster than my level of comfort. This resulted in a spectacular tumble. I limped back to the parking lot and tried to drive the car, but the pain in my right arm made it too difficult to change gears, so Pam did the gear shifting while I depressed the clutch. The next morning I went to the hospital, where they confirmed I had dislocated my shoulder. Not only that, I had fractured the head of my humerus! That's what happens when you let pride get the better of you.

I officially retired from the navy in 1989. Although I had not been involved in naval work since becoming an astronaut, severing my ties with the navy, the profession I had first embraced, was a significant moment in my life. I have always thought that the title of historian George Stanley's book *Canada's Soldiers: The Military History of an Unmilitary People* said it all. Canadians are not an aggressive people. And yet we have taken up arms and made important sacrifices when we believed it was necessary. Joining the navy had been my childhood dream. It was a way for me to serve my country, as my father and his father had done.

My takeaway from those years was that Canada was a peaceful country, but that in order to ensure that peace, it needed to maintain a credible Armed Forces, not only to ensure its own security and sovereignty but also to contribute to global security through its alliances. It was not acceptable to leave the heavy lifting to others. In other words, if Canada wanted to be viewed by its allies as a dependable partner, it needed to commit the necessary resources.

Unfortunately, despite the quality of its men and women, the navy was not, in my opinion, adequately supported during the time I served, or, for that matter, since then. With apologies to Theodore Roosevelt, Canada spoke softly (a good thing) but also carried a small stick. This would become painfully obvious to me later on, during my time in politics, particularly as foreign affairs minister, where sovereignty and collective security issues were of central importance to our foreign policy. Successive governments have failed to recognize that our ability to act in a credible manner on matters of international security, where we want to be seen as a useful contributor, requires more than words and goodwill; it requires the military assets to back up our words.

Skipping ahead to the present day, what has changed since my time in the navy? As we all know, Canada is bordered on three sides by oceans. Traditionally, we have focused most of our attention on the Atlantic and our membership in NATO. Now, with our recent, and necessary, focus on the Indo-Pacific region and the increased accessibility of the Arctic Ocean, Canada must urgently examine what its navy requires to ensure both our sovereignty and our global security interests.

In the end, governments decide how to apportion limited budget resources and then must live with the consequences. It's a choice. What governments can't do is have it both ways. If they choose to spend less on their armed forces, they cannot expect their allies to give their words more weight than they deserve. In my opinion, Canada needs a larger and more capable navy.

While the Astronaut Office was preparing for the next two Canadians to fly, several other space program initiatives were underway in Canada. Chief among them was the decision to consolidate the majority of Canada's space activities under one organization (including the astronaut office). This made eminent sense and led to the creation in 1990 of the Canadian Space Agency.

In 1992, given our decision to participate in the International Space Station program, a competition got underway to select four more Canadian astronauts. More than five thousand people applied. Being on the selection board, I can confirm that choosing the final four was challenging. Two of those who made the final twenty had actually been finalists with me back in 1983! One of the four we chose subsequently changed his mind at the last minute (before names were made public), opening the door for another candidate. In the end, we chose Dafydd Williams, Julie Payette, Chris Hadfield, and Michael McKay.

The year 1992 was an exceptional one for the Canadian astronaut program. In addition to scheduled flights for Roberta Bondar and

Steve MacLean, and the hiring of new astronauts, NASA also announced that Canadians would be invited to train as mission specialists. If we were to contribute hardware to the International Space Station, it stood to reason that we should also contribute crew. Those crew members would require the necessary knowledge and skills to help build, operate, and maintain the station. Consequently, they would have to be trained as mission specialists just like their American counterparts. Our other partners in the ISS negotiated the same deal for their astronauts. This would significantly influence my own future.

The year before, I had begun to experience numbness, tingling, and pain in my left arm. This worried me because I was hoping to have another opportunity to fly and, based on my knowledge of NASA medical selection criteria, I would not be considered "fit to fly" with such a condition. I also didn't know what was causing it. I consulted a neurologist, who recommended a CAT scan, and this led to the discovery of disc compressions in my neck, between two separate pairs of vertebrae. Years of playing football and squash had taken their toll. I was referred to a neurosurgeon at Sunnybrook Hospital in Toronto who was willing to perform the necessary surgery, after methodically warning me of all the risks that the procedure involved—even permanent paralysis or death! Nevertheless, I chose to go ahead, wanting to get rid of the numbness and hoping to fly again. The procedure involved accessing my neck vertebrae from the front and inserting cow bone spacers between the two pairs of vertebrae to relieve the pressure on the compressed nerves. The hardest part of the procedure was the intubation required while I was still conscious. That's when I discovered the power of the human gag reflex!

When I awoke from the operation, I was told it had gone well. For starters, I was alive and was not permanently paralyzed! My surgeon had warned me that the range of motion in my neck might be reduced, particularly when I tried to look behind me, but I never really noticed much of a restriction after the first few months, and I was certainly relieved that the numbness, tingling, and pain had disappeared. There were, of course, the usual family jokes about me now being part human, part cow. I had to put up with a lot of mooing!

Not so long ago, I would not have been able to benefit from such world-class surgery. My left arm would have become increasingly numb and weak, with no hope of recovery. Fortunately, a procedure was now available to correct the problem. Yes, it carried a small degree of risk, but after speaking with the neurosurgeon, I felt it was a risk worth taking. To this day, I'm glad I went ahead with the procedure.

When Roberta Bondar launched in January 1992, I flew down to Florida to provide live media coverage. It was an exciting moment, with all of Canada watching. I also asked Pam to join me, because I wanted to pop the question. Our relationship had grown serious and, should she be willing, I was ready to make the big commitment. We were staying in Cocoa Beach and my plan was to propose to her after Roberta's launch.

Based on our mutual recollection, here's what happened. We set off for what turned out to be a long walk on the beach and I began to talk about my current situation, perhaps to set the stage for my proposal and perhaps because I was a little nervous. I spoke about being a father with two teenagers, all of which Pam, of course, already knew, and that I wanted a more settled life with a partner, especially since some changes, including a move, might be coming soon. Perhaps I wasn't totally clear, but I'm relatively certain I said I hoped she might be willing to be that partner and to share her life with me.

As it turned out, Pam wasn't sure whether I had proposed, and after our walk, while I was taking a shower, she called her sister Heather and relayed our conversation. Heather confirmed that I had indeed proposed!

Another reason played into the timing of my marriage proposal. By then I knew that Canada was planning to accept NASA's invitation for two Canadians to train as mission specialists, and that their training would begin that summer at the Johnson Space Center. If I was chosen, this would mean moving to Houston, so I not only hoped to get married beforehand, but I also wanted Pam to know that if she agreed to marry me, we might be moving to Texas that summer.

My first spaceflight had been an extraordinary experience and I had always hoped that I would have another chance to fly. Space had worked its magic on me, and still does. Not only that, but in the nearly eight years since my flight, I had thought of many things I'd wished I'd had the opportunity to do on that flight, such as observing more closely the changes taking place on Earth because of human activity. A second flight might offer me that opportunity.

Given that many of my fellow astronauts were still waiting for their turn, and that there were only so many opportunities, I had kept my hopes in check. That spring, however, senior management at the Canadian Space Agency decided that Chris Hadfield and I would move down to Houston to train as mission specialists. This triggered a sequence of events that culminated in my marriage to Pam on July 16, followed two weeks later by our move to Texas.

Yves and Simone were delighted to hear that Pam and I were getting married. They had been hoping for quite some time that we would tie the knot, although I have to admit they weren't ready for the move to Texas. Most sixteen-year-olds form strong attachments to friends, and having to say goodbye to them caught them off guard. Fortunately, they chose to be positive about it, not to mention that the United States, and Texas specifically, had a certain lure for teenagers that Ottawa probably doesn't have. Don't get me wrong. I love Ottawa, but it can be a little sleepy.

Pam, of course, had her own furniture and other possessions, so we had to decide what to combine, what to sell, and what to put in storage. In addition, I decided to put my house on the market. It was a hectic time for both of us as we prepared for our new life together.

The wedding, at St. Columba Anglican church in Manor Park— where my children had sung in the choir and I had been the rector's deputy warden—was a modest but elegant ceremony, presided over by Father Ralph, followed by a lovely reception given by Pam's parents,

Diana and Jim Soame. We even managed a short honeymoon at my parents' cottage on the Gatineau River before I flew to Houston to find a house and a school for Simone and Yves, while Pam took care of the move. The plan was for them to join me in early August at our new home and to be ready for school, which in Texas started in mid-August.

With the help of the Canadian consulate in Dallas, we found a home in Clear Lake, ten minutes from JSC, and I arranged for Yves and Simone to attend Clear Lake High School, a school of three thousand students that proudly announced at its front door: "We are a gun-free, drug-free school."

Welcome to America!

SIX

ALTHOUGH WE DIDN'T KNOW IT AT THE TIME, Pam and I were destined to live in Houston for more than eight years.

Let me just say this: Don't believe everything you hear about Texas and Texans. We thoroughly enjoyed our life in the Lone Star State. As we settled in, we began to explore our neighbourhood of Clear Lake, equidistant from Houston and Galveston. It was suburbia, but it suited us well and was just a short drive from the Johnson Space Center. Our neighbours quickly welcomed us, and we discovered first-hand that Texans are warm and generous. We lived in a predominantly Republican neighbourhood, whose residents I would describe as devout Christians and patriotic Americans. Several approached us to invite us to their local church. Being progressive liberals and lapsed Christians, we tried to steer clear of religion and politics and not stick out like sore thumbs.

Yves and Simone entered grade 11 at Clear Lake High School. Although they had gone to French school in Canada, they spoke mostly English at home and were very comfortable following the Texas high-school curriculum.

Meanwhile, I began my training with Astronaut Class 14, known as "The Hogs." Our group of twenty-four included nineteen Americans, two Canadians (Chris Hadfield and me), two Europeans (one French and one Italian), and one Japanese. Four of the Americans would train as pilot astronauts and the rest of us as mission specialists. I was the only one who had already flown.

Our training got underway right away and kept us busy. We spent a weekend getting to know each other while doing survival training in the bush in the state of Washington. It was an opportunity to refresh the skills I had learned as a scout, and I have to admit that "surviving" outdoors at that time of the year was not terribly difficult.

Over the course of the next year, we would get to know NASA, visiting many of its centres and facilities, some of which focused on space, others on aeronautics. I could not help but be impressed by the size and breadth of this organization. With a current workforce of some eighteen thousand, an annual budget now exceeding US$25 billion, and almost twenty facilities spread across the U.S., NASA (to use its own words) "explores the unknown in air and space, innovates for the benefit of humanity, and inspires the world through discovery"— an enviable and totally justified claim. NASA is unquestionably a world leader in whatever it undertakes, and I was more than proud to be working at the Johnson Space Center.

We began classroom lectures almost immediately and were introduced to JSC's many training simulators. These were high-fidelity fixed and mobile-base simulators that allowed us to practise the launch, on-orbit, and re-entry phases of flight, while training staff introduced malfunctions, to which we had to respond. Robotics simulators allowed us to capture, move, or release a payload using the Canadarm. EVA (spacewalk) training was done underwater, initially in a large swimming pool known as the WETF, or Weightless Environment Training Facility, and later in an even larger pool known as the NBL, short for Neutral Buoyancy Laboratory, which could accommodate mock-ups of portions of the ISS. (As you may have noticed, NASA is awash in acronyms! For example, EVA stands for extravehicular activity, a rather bureaucratic way of describing a spacewalk.)

We also began training in the T-38 Talon jet, a new and exhilarating experience for me. Although capable of supersonic flight, the aircraft was flown at high sub-sonic speeds to avoid breaking the sound barrier and creating a sonic boom. Although it was not a selection requirement for any of the mission specialists to be pilots, I suspect most, like me, already had a pilot's licence of some kind. We flew in the back seat of this two-seat jet and supported the pilot astronaut in the front. Most of the pilots allowed you to take the controls from the back seat, except for during take-off and landing, and you shared the communication and navigation duties.

Sometimes, T-38s would also fly in formation. I once flew with astronaut Scott Altman, the pilot who had done the flight scenes for Tom Cruise in the movie *Top Gun*, and he let me take the controls from the back seat. Given how close the aircraft are to each other when in formation, this required my total concentration.

Because we flew in the T-38s regularly, we had to learn how to eject in case of a major malfunction, including how to land under a parachute. Our training for this took place on land and consisted of our parachute being opened behind us and then being raised into the air by a rope, somewhat like a kite. When the rope was released, we descended under an open canopy and experienced a landing. We did the same from the back of a ship and landed in the water. Having previously skydived in Canada, I was already familiar with free-falling and opening my parachute, as well as with the experience of landing on the ground, although knowing how to get out from under a canopy after a water landing was a new experience.

Soon after our new class met, we began to organize social events so that our families could get to know each other as well. Barbecues and pool parties were favourite weekend activities. Creating strong links is important because of the need to support each other, not just the astronauts but our families. When a crew launches, it can be a stressful time for those left behind, perhaps becoming more so in the years after the *Challenger* disaster.

As an early demonstration of team building, I co-volunteered to lead the Hogs in the annual chili cookoff at JSC and was given the green light to prepare my world-famous chili (the "world" consisting of my immediate family and tens of other people), only to discover that Texans, or at least the ones doing the judging at JSC, frowned on having beans and tomatoes in their chili. An authentic Texas chili did not countenance such ingredients. Fortunately, my fellow astronauts, none of whom was Texas-born, forgave me the oversight.

It takes about a year for a new class to complete "basic" training, and graduating is an important milestone. Even though I had already flown, mission specialist training was as new to me as it was to my classmates. In some ways I felt an additional pressure to perform well. Fortunately, we all graduated and received our silver astronaut pin, meaning we were now eligible for a flight assignment. We had joined the end of the queue and would now support the Astronaut Office in different ways until it was our turn to fly.

A bunch of us celebrated by flying to Grand Cayman for several days of scuba diving. Pam joined me, having completed a scuba course in Houston. It was a wonderful holiday, which included several day-time dives, one night dive, and swimming with stingrays at Stingray Cove. If you weren't too nervous, you could feed these large fish by holding food in your open hand (flat and extended upwards) as one of them swept over it and vacuumed it up with its mouth.

It's a tradition in the Astronaut Office that the newest class is responsible for the annual Christmas show. Chris Hadfield and I, inspired by comedians Martin Short, Christopher Guest, and Harry Shearer, choreographed and executed a memorable (some might say legendary!) "Canadian men's pair synchronized swimming" routine to the finale of Beethoven's Ninth Symphony. To my knowledge, only a few copies of this video remain, or I hope that's the case, and I have one of them. The vault can be unsealed when I'm gone.

As a Canadian living in Houston, I took every opportunity to highlight moments of Canadian pride to my American counterparts. This included when the Blue Jays won the World Series for the second time in 1993. We had finished watching the final game when some-one suggested we raise the Canadian flag at JSC for all to see. Pam and I thought this was a good idea and it so happened that a flag was available. We hatched a plot to raise it on one of the three tall flagpoles in front of JSC's main administration building. Maximum stealth would be required since JSC security guards regularly patrolled the site and the flags themselves would be illuminated by large spotlights.

We entered JSC and drove to the administration building. We switched off the engine, got out, and approached one of the flagpoles. The pole we chose was flying a huge NASA flag that almost smothered us when we lowered it to the ground. We clipped on our small Canadian flag and hoisted both up, all the while nervously watching for security guards. Fortunately, our act went unnoticed until the following morning, when the Canadian flag, despite its modest size, was visible for all to see.

It wasn't long before I received a call, asking me whether I knew anything about a Canadian flag on a JSC flagpole. Given that there weren't many Canadians at JSC, the detective work to find the culprit was not that challenging. I confessed to being the guilty party and, while I was at it, sold out my accomplice, Pam. Predictably, our audacious act was received with good humour by all, including the centre's director.

Until you're named to a mission, you provide support to the Astronaut Office. In fact, most of an astronaut's career is spent on the ground in support roles of one kind or another. You might be a "Cape Crusader," requiring you to spend time at the Kennedy Space Center in support of crews getting ready to fly. You might be a capcom in Mission Control, the person who speaks to the crew during a mission. You might be assigned to support robotics or EVA training or to evaluate new procedures or space hardware. All these assignments add to your experience and allow the Astronaut Office to continuously assess you. In the case of mission specialists, this includes seeing if you have strong skills as a spacewalker or robotic arm operator.

Meanwhile, of course, you are hoping for that special phone call from the head of the Astronaut Office informing you that you've been assigned to a flight. Eventually, someone from your class does get named, and this gives you hope that your turn will follow shortly, although nothing is guaranteed. Such is the life of an astronaut: there is truly never a dull moment.

One of the "firsts" that I was most proud of in my astronaut career was being chosen as the first non-American to become a capcom. Only astronauts can become capcoms. The astronauts on the Mercury program, NASA's first manned space program, had insisted that the person they spoke with in Mission Control during a mission had to be an astronaut, because only an astronaut truly understood what they were going through. Capcom was short for capsule communicator, a holdover from those early days. I believe that my already having been to space was a determining factor in my selection for this job, despite my not having flown as a mission specialist. With spaceflight experience, I did not have to imagine what the conditions were like: I had been there. I felt honoured to be given this role.

To get up to speed, I initially sat beside experienced capcoms and watched them perform their duties. When I was ready to do the job myself, I was assigned to a specific mission and remained with that mission from beginning to end: from the moment the crew began its training until it returned from space. This allowed me to get to know the different crew members and become familiar with the details of the mission. The first flight I capcomed was STS-65, a fifteen-day International Microgravity Laboratory mission that flew on *Columbia* in July 1994. Over time, I was involved in seventeen missions, either as a trainee or as the actual capcom. Other Canadian astronauts would eventually serve in the same role.

Mission Control is the nerve centre from which NASA directs space missions. It is under the control of one person, the flight director, who is supported by a small army of flight controllers, some sitting in Mission Control, others supporting from back rooms. Essentially, all of NASA's resources are at the disposal of the flight director. As capcom, I relayed his or her instructions to the crew and responded when the crew called Mission Control ("Houston" for short). For situational awareness, I also monitored everything that was being said by the flight controllers in Mission Control.

Typically, a mission would have three capcoms for the on-orbit phase of the flight, each taking an eight-hour shift, with some overlap for handover purposes. Pilot astronauts would act as capcoms for the much shorter but extremely dynamic launch and entry phases. A good capcom must be clear, quick, and concise and come across as calm and reassuring, no matter what. The crew needs to feel that the situation is under control. I should point out that Mission Control has access to a mountain of data, much more than what the crew sees.

A highlight of my time as a capcom was the transition from the original Mission Control Center to the one used today. I had the opportunity to participate in some of the equipment evaluations as the new Mission Control was brought on line in 1998, in time for the International Space Station program.

The original Mission Control Center is now a National Historic Landmark. Becoming operational in 1965, it was from this location that the last nine Gemini missions (Gemini 4 to 12), all of the Apollo and Skylab missions, and all shuttle missions until 1998 (including my own first flight) were controlled. Before that, the Mercury missions and the first three Gemini missions, two of which were unmanned, were controlled from Cape Canaveral.

Any member of the public visiting the original Mission Control cannot help but notice how primitive its equipment consoles and controls are compared with those in the current facility. And yet in its heyday this was state-of-the-art technology. What the visiting public will not see are the incredible flight control teams that sat at those consoles and made it all work.

So much history was witnessed and indeed made in this location, including the first landing on the Moon. I can still vividly remember when, after a period of nerve-racking and seemingly interminable silence as the lunar module descended, Neil Armstrong announced to the world: "Houston, Tranquility Base here. The *Eagle* has landed." It was also from here that the dramatic rescue of Apollo 13 unfolded, and Flight Director Gene Kranz uttered the words: "Failure is not an option."

Training astronauts was a process that JSC had fine-tuned over decades, beginning with the first-ever group of astronauts, the Mercury Seven. When something works well, you don't change it unless you're sure you have something better to take its place. That said, there were occasional challenges to accepted practices, some for good reason, others possibly motivated by the desire to cut costs. For example, while flying in T-38 jets was considered an integral part of training, the question was sometimes asked by the bean counters: Is it really necessary for astronauts to fly in a jet when they're training to fly in a space shuttle? Isn't simulator training sufficient? The answer was a no-brainer: flying allows astronauts to stay sharp when it comes to thinking on their feet, communicating quickly and accurately, and reacting to the unexpected—the same skills required for spaceflight.

The NASA T-38 jets were based at Ellington Field, fifteen minutes from where I lived, and while I flew in them like everyone else, I also wanted to maintain my private pilot proficiency. Fortunately, Ellington Field had a civilian flying club with aircraft available for rent, including Cessnas and Pipers. This allowed me to fly regularly, and I would take Pam up to explore different parts of Texas. We would typically fly to a place of interest, land, have lunch, and fly back home.

After settling into our new home, Pam and I began exploring what Houston had to offer and were delighted to discover this included two of our favourite activities, theatre and classical music. Along with a few other astronauts from our class, we took out subscriptions to both.

After completing grade 11, Yves announced that he wanted to return to Ottawa to finish his studies. He was now seventeen, and we supported his decision. He wanted to spread his wings, something to which I could relate, and would be living on his own, although close to my parents, who would be there to lend a hand. To our delight, Simone stayed with us for another year and then she too left to begin engineering studies at Queen's University in Kingston, Ontario.

In 1993, Pam got back into nursing, starting with a day surgery in nearby Dickenson. A year later, she switched to Planned Parenthood in downtown Houston, a potentially hazardous job in a relatively anti-abortion, evangelical-minded state. I used to say that, between me being an astronaut and Pam working at Planned Parenthood, there was no contest: Pam had the more dangerous job. In fact, security personnel were posted in front of her building most of the time, as protesters showed up regularly. I supported Pam's work without reservation, but I was always secretly relieved to see her arrive home safely after work.

With Simone's departure, Pam and I had become empty nesters and decided we needed a dog. Whereas Pam had grown up with them, this would be a new experience for me. And so in March 1995, a twelve-week-old puppy, Rosie the Bouvier, came into our lives. Her full name was Texas Arundel Alamo Rose. She was fawn-coloured rather than the more common black or brindle. She eventually grew to over a hundred pounds. She was calm and protective by nature, and she brought us much joy. Her presence gave me peace of mind for those occasions when I was away.

Not long after acquiring Rosie, Pam and I made the biggest decision of all—to start a new family. As is often the case when couples decide to have children, it doesn't happen right away. After a miscarriage and further attempts, we decided to look into fertility treatments. Those who have been through such treatments know it's emotionally difficult to repeatedly cycle from hope to disappointment. After many attempts, we decided to let nature take its course.

In the fall of 1995, I got the news that my brother Braun had returned from working as a nurse in Saudi Arabia and had been diagnosed with non-Hodgkin's lymphoma. I had trouble believing it for many reasons, not the least of which was because Braun was one of the fittest people I knew. Unfortunately, being fit didn't mean you couldn't get cancer. He began a course of chemotherapy, and we all held our breath.

———

As I continued to perform the job of capcom, some of my classmates began receiving flight assignments. This was encouraging, and I was hopeful that my turn would soon come. I secretly hoped for a mission in which I could operate the Canadarm.

Crew selection is done by the head of NASA's Astronaut Office and is driven primarily by the requirements of each mission. There may be a need for spacewalks, robotic operations, or rendezvous operations, or the focus may be on science experiments. In some cases, prior flight experience may be a prerequisite, weighed against the need to fly new people to ensure a steady supply of experienced astronauts. Crew chemistry is also a factor, although probably more important in the case of long-duration missions. Finally, while all astronauts belong to the same pool, NASA must also ensure that foreign astronauts fly at frequencies that are tied to international agreements.

Is the competition to be chosen for a mission intense? You bet! Astronauts all have type A personalities. That being said, I never once heard anyone from my class complain they were being ignored or bypassed.

My call from the head of the Astronaut Office came in the fall of 1995. I was named to mission STS-77, scheduled for liftoff the following May, and I could not have been more thrilled. Almost twelve years after my first flight, I would be returning to space and this time as a mission specialist. It was cause for celebration, particularly since the previous months had weighed heavily on me, with Pam and me trying but not succeeding in having a baby, and Braun being diagnosed with cancer. My assignment to a mission came at just the right time.

I would be flying on *Endeavour* with five crewmates: Commander John Casper, Pilot Curtis Brown, and Mission Specialists Andrew Thomas, Daniel Bursch, and Mario Runco, Jr. I would have two primary responsibilities. First, I would be operating the Canadarm (as I had hoped) and capturing a spacecraft called Spartan 207 after it was deployed by Mario. This spacecraft would be carrying a large folded antenna, made of Mylar plastic, that would be inflated to full size at a safe distance from the orbiter. Measurement data would then be collected, and the antenna would subsequently be detached from Spartan and re-enter Earth's atmosphere a few days later. Following this, we would rendezvous with Spartan, and I would capture it with the Canadarm and restow it in the payload bay.

Andrew Thomas and I would also be activating a large pressurized module called Spacehab, which carried scientific experiments to be performed throughout the mission. Spacehab was located in the payload bay and was accessible from the mid-deck through a short tunnel section.

Performing experiments during our mission required us to meet with the people who had designed them, to fully understand the underlying science and the procedures we would have to follow. The crew have to feel comfortable with the experiments they will conduct, since those who designed them obviously won't be on orbit to assist, and there is no guarantee we can communicate with them in real time. We need to assume we're on our own. We're also conscious that each experiment represents years of work, and we don't want to make any mistakes.

Learning about the experiments meant flying to places across the country to receive the necessary training. For example, I flew to the University of Florida in Gainesville to be trained on the Commercial Float Zone Furnace experiment, which I would be performing during the flight. I also made trips to the Goddard Space Flight Center in Greenbelt, Maryland, which was responsible for the Inflatable Antenna experiment on the Spartan spacecraft.

As I was driving to the airport to take a flight to Washington, D.C., to meet with the Spartan engineers, I accidentally ran a red light and, despite slamming on the brakes, hit the side of a truck. Fortunately, no one was injured. I learned an important lesson that day: I may have thought I was a good driver, but I wasn't as good as I thought. The accident was clearly my fault. My mind was elsewhere when it should have been on the road. Years later, this incident would influence my approach to the serious issue of distracted driving when I became Canada's minister of transport. At the time, it was a wake-up call to stay focused on the task at hand, and to keep my ego in check.

In January 1996, Pam ran the Houston Marathon in the impressive time of three hours and thirty-six minutes. A few days later, she announced that she was pregnant! We were, of course, elated with the news. But life has a way of getting real very fast. I had hoped to invite Braun to my launch. Sadly, it was not to be. Despite his treatments, Braun died that winter. He was only forty-eight years old. I was fortunate to visit him a few days before his death and to say goodbye to him. I sat with him as he lay quietly in his hospital bed drifting in and out of consciousness. I held his hand and thought about the life we had shared. We had been best buddies from start to finish, brothers in arms throughout our childhood, getting into trouble together and later watching over each other as we travelled through Europe. He could make me laugh like no one else, even when I was mad at him. I loved him very much and still miss him. I spoke at his funeral.

Like the loss of Jacqueline nine years before, losing Braun jolted me. I had now lost two people with whom I had shared so many special moments in my life. More than anything, these losses made me realize the importance of family, including the family into which I was born. I had lost my older brother and I realized I didn't know much about my younger brothers. I had left home at sixteen, when Charles was eleven and Philippe was six. I hadn't seen much of them from that moment onwards. Now they were married adults with their own families. Braun's death made me resolve to see them more often when I returned to Canada.

NASA missions, whether crewed or uncrewed, have carried mementoes to space since the 1960s. When shuttle astronauts are assigned to a mission, they are allowed to take along a small number of personal objects, such as their spouse's wedding ring, a special coin, or something belonging to a relative or close friend. For example, on STS-77 I took with me a United Nations medal belonging to Gen. Roméo Dallaire. Roméo and I had known each other since our days at military college, and he was going through an extremely rough patch at the time, in the aftermath of the Rwandan genocide. I was later able to present the medal to him when he visited the Canadian Space Agency. I also flew a French and English copy of the Canadian Charter of Rights and Freedoms, which I consider to be Canada's most important document, and I subsequently presented them to Prime Minister Jean Chrétien.

As I was training for my flight, my Canadian colleague Bob Thirsk was also training for his, STS-78, due to lift off a month after us. Bob had always been a Bobby Orr fan and had offered to fly his number 4 Bruins jersey on his mission. When the jersey arrived at JSC, it was sealed in a special bag in preparation for forwarding to KSC for storage in the STS-78 orbiter. But just before it left JSC, my crewmate Mario Runco, a New Yorker and a serious Bobby Orr fan himself, got wind of the jersey's presence in our building. He proceeded to "borrow" it, remove it from its special bag, and bring it to our crew office, where I was the only one present. With great excitement, he asked me to photograph him holding the jersey. And then, unable to stop himself, he put it on and asked me to take another picture. I have rarely seen so much pleasure on someone's face.

As I described earlier, crews spend many months preparing for their mission, and STS-77 was no exception. Our crew trained in various simulators, practising what we would do on orbit. This, of course, required us to interact with Mission Control. To my delight, Chris Hadfield was one of the capcoms assigned to STS-77. As a Canadian, it was a particular pleasure for me to speak with him, not only during simulations but later during the mission itself, knowing that I was talking to a fellow Canadian. The two of us had trained together as

Hogs, with Chris flying first on mission STS-74 in November 1995. He would go on to achieve many firsts for Canada, including commanding the International Space Station.

Endeavour was set to lift off on a ten-day mission at 6:30 a.m. on May 19, 1996. The post–wake-up routine reminded me of my first flight almost twelve years before. I got up around two, shaved and dressed, and had my vitals checked. I then ate breakfast with the rest of the crew, with NASA cameras rolling. Then, with the assistance of a suit technician, I put on my bulky orange Launch and Entry Suit and verified that it worked properly. After thanking our NASA support staff, we headed out to the launch pad and took the elevator up to level 195, walked across a gangway leading to the orbiter side hatch entrance, and, with the assistance of NASA technicians, were helped into our seats.

Five things were different this time. The first was that I had experienced a launch, so it was not a trip into the unknown. I knew what to expect if all went according to plan. The physical experience would be the same and I was ready for it. It would be like taking your second ride on the same roller-coaster. The noise, vibration, and acceleration would be known quantities.

The second was that, having witnessed the *Challenger* disaster, I now knew, in a concrete way, that things could go wrong. That was part of the reality with which all crews had to deal, even though many safety improvements had been made and the shuttle had flown fifty-one successful missions since then.

The third was that I was now a mission specialist and, consequently, had a much greater knowledge of the vehicle in which I would be flying. I also knew my crew much better than on my first flight, having trained with them for many months. I was aware of the different malfunctions that could occur, and how the crew would respond. I also knew that a great deal of redundancy was built into the vehicle.

The fourth was that the orbital inclination of our mission was thirty-nine degrees, meaning that *Endeavour* would not be flying over any part of Canada, since the most northerly point in our orbit was at a lower latitude than Canada's most southerly location, Point Pelee, which sits at 42° North. It reminded me how fortunate I had been to fly at an inclination of fifty-seven degrees on my first flight.

Finally, the fifth difference was that the crew now wore the large orange pressure suits, designed in the aftermath of the *Challenger* disaster to ensure a greater degree of protection. Although they were not particularly comfortable to wear, they improved our chances of survival in certain potentially catastrophic scenarios. Fortunately, we didn't have to wear them on orbit and could don comfortable clothing (shorts and a polo shirt in my case) until it was time for re-entry.

As I waited for liftoff, I felt that same sense of excitement I had experienced on my first flight. And, yes, I felt those same butterflies. Most of all, though, I could hardly wait to get back to space. How would the launch feel? What would it be like once I reached orbit? My prayers were soon answered as the countdown proceeded without a hitch and *Endeavour* lifted off on time, headed for space.

It was every bit as wondrous as the first time as *Endeavour* rose into the morning sky and I experienced once again the euphoria of seeing Earth, our luminous planet, against the infinite darkness of space. Once again, I would know the incredible freedom of being weightless. I had long dreamed of this moment. My sense of gratitude was profound at having been given the opportunity to do this not once but twice.

On my first flight, as a payload specialist I had not been involved in transitioning the orbiter from the "ascent" configuration to the "on-orbit" configuration, as happens after main engine cutoff (MECO)—for example, disconnecting our seats from the mid-deck floor and stowing them until the re-entry. This time it would be different. Andy Thomas and I would also have to access the Spacehab module and activate it by connecting all the experiments to power and data cables.

This required getting down to work almost immediately after

MECO, but not before enjoying a few minutes looking out the window and then getting out of our bulky pressure suits, which would be stowed until it was time to return to Earth.

Apart from seeing Earth, what excited me the most about the mission was operating the Canadarm. On flight day 2, Mario Runco deployed the Spartan spacecraft and *Endeavour* manoeuvred away from it so that we could observe its giant inflatable antenna being deployed. What a spectacular sight that was! The parabolic antenna, once fully inflated, measured fourteen metres across and was supported by three twenty-eight-metre-long inflated struts. The purpose of the experiment was to explore the feasibility of using inflatable structures in space because of their obvious low weight and volume compared with conventional structures. As we watched the antenna inflate from a safe distance, the ground collected data. When this task was completed, the antenna was released from Spartan and would eventually burn up on re-entry.

We then approached Spartan so that I could retrieve it with the Canadarm. The end of the arm has a mechanism called the end effector, which can capture a spacecraft that is equipped with a special grapple fixture, essentially a peg protruding from a circular plate mounted on the spacecraft. This is a delicate operation. Whether you're grappling a spacecraft stowed in the payload bay or one that is free-flying, the same capture operations are required. That being said, there is more at stake in the latter case since a mistake in capturing a free-flyer can cause it to start rotating or backing away or both, in some unpredictable fashion, making it difficult to recapture unless it can be stabilized.

Endeavour had been positioned perfectly with Spartan within reach of the arm, and there was no perceptible relative motion between it and the spacecraft as both travelled through space at twenty-eight thousand kilometres per hour. I now had the green light to move the end effector over the grapple fixture and capture the spacecraft. It all went according to plan, and I restowed Spartan in the payload bay by lowering it into a docking mechanism equipped with four latches.

My other main responsibility, after helping set up the Spacehab module, was to perform several of the science experiments it contained. This was exciting work—leading-edge science investigating how weightlessness produced different outcomes than on Earth, where gravity is the dominant force.

One experiment in particular fascinated me. It was called a Commercial Float Zone Furnace and was used to melt semiconductor material and subsequently cool it until it resolidified. I performed this experiment over several days, melting and cooling a variety of materials. After the flight, the scientist who had designed the experiment would have the opportunity to examine whether the crystal structure of the resolidified materials had been altered in such a manner as to improve their semiconductor properties.

Once again on this mission, I was allowed to bring a selection of music with me, this time on CDs. I had been pleased with my choice of baroque music on my first flight but decided that this time I would also bring some jazz: Miles Davis, Diana Krall, John Coltrane, and other favourites of mine, music I enjoyed after a hard day's work when I was in a reflective mood. My personal verdict, based on a sample of one? Although ordinarily I love listening to jazz, it doesn't really go with space, unlike classical music. Jazz is meant to be enjoyed on Earth! You heard it here.

While on orbit, I was greeted with an unexpected surprise. Pam was roughly five months pregnant, and her first ultrasound occurred after we launched. Because we communicated daily by email, she asked me if I wanted to know the sex of our baby. I told her I preferred to keep it a secret, but my resolve didn't last, especially since every day she offered to tell me, and eventually I gave in. I learned on orbit that we were expecting a boy. We would name him Adrien Braun Garneau.

As was the case on STS-41G, the mission went by far too quickly. Ten days after launch, with all our objectives met, it was time to come home. This meant configuring the orbiter for re-entry: stowing the Canadarm, closing the payload bay doors, deactivating the Spacehab module, securing loose objects, setting up our seats, donning our orange Launch and Entry Suits, and performing the deorbit burn. It all went according to plan, and *Endeavour* touched down at KSC at 7:09 a.m. on May 29, the culmination of a successful mission.

At this point we would normally exit *Endeavour* and do a walkaround to inspect it, but not in this case. We still had one more scheduled task to complete: Detailed Supplementary Objective 331, whose purpose was to measure our ability to get out of the orbiter quickly and run several hundred yards from it while still fully dressed in our bulky orange flight suits, essentially simulating an emergency evacuation of the orbiter such as might occur after landing if, for example, a toxic propellant had been released from the vehicle. To accomplish this, a special room had been wheeled up to the orbiter, equipped with a treadmill on which we would each run for several minutes while technicians measured our energy expenditure. Performing this demanding test immediately after landing while still fully suited made me realize just why it's so important for astronauts to exercise on orbit to minimize deconditioning, advice I had taken seriously.

After our postflight debriefs, I took time off so that Pam and I could focus on the big day, the arrival of our son Adrien, in October. Although I was already a father, twenty years had gone by and things had changed, some would say at breakneck speed. The world awaiting Adrien would be quite different from the one that greeted Simone and Yves, and I knew I would have to adapt my parenting skills accordingly.

Pam had found a good ob-gyn and was having excellent monthly checkups. The two of us took a class so that I could be present at the delivery and offer encouragement. At forty-seven, I was by far the oldest father in the class, and definitely the only astronaut. At her last checkup in October, which happened close to Halloween, Pam painted a big pumpkin on her belly, and when the doctor lifted her blouse to examine her, he burst out laughing. I don't think anyone had ever surprised him like that.

Pam checked in to the hospital on the morning of October 24 for an induced delivery. We had celebrated the night before with dinner at Frenchies, our favourite Italian restaurant, ordering pasta pappardelle in a creamy tomato sauce. The next morning, as planned, labour was induced. Dilation progressed normally at first, but then stopped prematurely. By this time it was evening, and the doctor decided to perform a caesarean. I was allowed to be present for the delivery, which went smoothly. Adrien entered our lives a healthy eight pounds thirteen ounces. He was quickly reunited with his exhausted mother.

For me, 1996 had been a year of both sorrow and joy: sorrow at losing my dear brother Braun and joy at welcoming Adrien into our family. The circle of life. It had also been a year when my dream of returning to space had come true. I was now a mission specialist with flight experience.

SEVEN

AS I RESUMED MY DUTIES IN THE ASTRONAUT OFFICE, an ambitious new program was taking shape: the International Space Station. Each of the partner countries would contribute hardware and share in the costs of operating and maintaining this permanently inhabited orbiting facility dedicated to scientific research in microgravity. Canada would contribute the Mobile Servicing System, which included the large Canadarm2 robotic arm and Dextre, a two-armed robot capable of performing fine-control tasks such as module replacements, in lieu of having to send astronauts out on spacewalks.

The challenge was to determine the optimum assembly sequence, bearing in mind the equally important need for ongoing resupply missions. Canadian robotics, on both the shuttle and the station, would play a critical role during assembly, a process that would take about ten years and thirty missions to complete. Not only was the sequencing of the missions critical, but the physical interfaces between the different pieces of the station had to be such that it all fit together. Over time, the station would grow and incorporate new capabilities. The first piece would be the Russian *Zarya* module, to be launched in November 1998.

One of the objectives was to allow a small crew, known as an expedition crew, to live on the station as soon as possible, well before station assembly was complete. Typically, expedition crews would spend about six months on orbit, operating and maintaining the station and performing scientific experiments as new elements were

added. Periodically, there would be visits from Soyuz vehicles bringing up a new crew or shuttles ferrying up new pieces of the station.

In preparation for the station being inhabited, it was necessary to
plan for the resupply missions that would bring essentials to the
crew as well as new scientific experiments for them to perform. This
would be done by unmanned spacecraft such as the Russian Progress
cargo vehicle, which could dock itself to the station. Later, the
European Space Agency would also perform resupply missions, using
an unmanned Automated Transfer Vehicle launched on an Ariane
rocket from Kourou, French Guiana, and also capable of docking
itself to the station. Similarly, the Japanese space agency would provide unmanned H-II Transfer Vehicles launched from Tanegashima,
Japan, which would rendezvous with the ISS and be grappled by the
Canadarm2 robotic arm and docked.

Because Russia was a partner in the ISS and would contribute
cosmonauts to every expedition crew, it became necessary for all
astronauts assigned to an ISS mission to acquire some proficiency in
the Russian language. It also meant going to Star City, the once-secret
cosmonaut training facility outside Moscow. For obvious reasons,
the astronauts who would fly to the station on a Soyuz rocket (or
return to Earth in a Soyuz module) had to achieve a higher level of
fluency. Consequently, many astronauts began learning Russian. In
due course, I would receive about two hundred hours of instruction
and visit Star City.

Meanwhile, on the home front, Yves had moved to the skiing
resort of Verbier in the French part of the Swiss Alps to indulge his
love of skiing and snowboarding. Eventually he would settle there,
become a professional photographer, get married, and become a
Swiss citizen. On Christmas Day 1997, I received a rather groggy call
from him, telling me he was in hospital with a broken back from a
snowboarding accident (his exact words). My heart nearly stopped.
What he really meant to say was that he had fractured a vertebra in

his lower back and was now immobilized and sedated, but not paralyzed. I was on a plane to Geneva the next day, followed by a train ride to Martigny, where he was hospitalized. A kind Swiss gentleman saw me looking for a taxi at the train station and insisted on driving me to the hospital, where I was quickly admitted to Yves' room. He was in bed, "wearing" a metal frame to immobilize most of his body. It reminded me of my own skiing accident when I was eight years old and spent two months in hospital with my left leg in traction. (The apple doesn't fall far from the tree. Literally!) That night, I breathed a sigh of relief. Although it would take months for Yves to heal, he would make a full recovery.

Parenting has its share of heart-stopping moments, and this was certainly one of them. Like me, Yves was a bit of a thrill-seeker and did not hold back. I could not help but reflect on the risks I had taken in my own youth and the fact that I had not been fully aware of the anxiety I must have caused my parents. Perhaps it was only fair that I was now learning what I had put them through.

As parents, we often wonder what similarities our children share with us. In Simone's case, it was her analytical approach to problem-solving. She was rigorously methodical in everything she tackled. For that reason, I was not surprised that she chose to study engineering, a discipline that even today attracts too few women. She graduated from Queen's University in the spring of 1998 and went on to do her master's degree at the International Space University in Strasbourg. Although I have never consciously pushed my children along a particular career path, my own space career more than likely influenced her decisions. It was in Strasbourg that she met her future husband, Ozgur Gurtuna, a Turkish student also doing a master's. Following that, she spent three months at the NASA Jet Propulsion Laboratory in Pasadena, California, and then a year at the Canadian Space Agency. For a while, it looked as though she might be embarking on a career in the planetary sciences, but then she did a hard pivot to the film business, doing research for documentaries and film scripts, some of which were space-related, and discovered that she had a talent for writing. Since then, she has continued to hone those skills.

While Yves and Simone were getting on with their interesting lives, I continued to do the job I loved. At the same time, Pam and I were beginning to do some long-term thinking, imagining the day when we would return to Canada, although hopefully not before I was assigned to an ISS mission. I was forty-eight years old and felt that my flight prospects were good, provided I remained healthy.

It goes without saying that astronauts must be free of medical problems if they are to fly. To ensure this, NASA subjects them to frequent exams. When these take place, there is always the risk of discovering something that could disqualify an astronaut, at least temporarily, from a flight assignment. To minimize the chances of that happening, astronauts focus a great deal on physical fitness, not unlike athletes, although with somewhat less intensity and focus. Most exercise daily, and I was no exception. I concentrated mostly on my cardiovascular system: playing squash, running on a treadmill, or riding stationary bikes. I also lifted weights and used various machines to maintain muscle strength.

I also went to the gym on Sunday afternoons for a light workout. One time, as I entered the gym, I noticed the lights were out except in the weightlifting corner. Someone was on his back bench-pressing. I didn't recognize him as one of the usual Sunday crowd. I wandered over to see who it was and recognized John Glenn. Here I was, looking down at one of the icons of the space age. (I loved how he had been portrayed in the film *The Right Stuff*.) I introduced myself and we chatted. He was most obliging. He then resumed his workout. He was seventy-seven years old at the time, and still a great inspiration.

Glenn became the first American to orbit Earth on February 20, 1962, aboard the *Friendship 7* capsule. Following his historic flight, he remained an astronaut for two more years before resigning, not having had a second opportunity to fly. As we all know, he entered political life, becoming a senator. For Americans, John Glenn embodied the qualities they most admired: he was not only a marine who didn't smoke or drink; he was a family man and a model of God-fearing patriotism.

Nobody thought he would ever fly again until NASA announced his assignment as a payload specialist on shuttle flight STS-95, scheduled to lift off on October 29, 1998. He would perform health science experiments, focusing on how weightlessness affected older humans. I was one of many astronauts sent to KSC to cover his launch and provide media commentary. I had never seen so much excitement for a shuttle flight. John Glenn was returning to space!

I bring this up for a reason: What struck me about Glenn's flight was the high level of interest among older people. They were everywhere, and their pride was showing. You heard it in every interview. Yes, John Glenn was going to fly again, but equally important, a seventy-seven-year-old was going to space. Life, after all, didn't end when you turned sixty-five; perhaps another great leap for mankind!

As you have gathered by now, even astronauts have their heroes. Glenn was certainly one of mine. I would meet another one after my return to Canada in 2001: U.S. astronaut Jim Lovell. As many will know (particularly those who saw the movie), Lovell was the commander of Apollo 13, which experienced a life-threatening malfunction on its way to the Moon. For me, he was the embodiment of the individual who radiates competence, keeping it together under extreme pressure.

Meeting him was a truly memorable moment in my life. The two of us had dinner together, and I was able to get to know him much better than is usually the case when there's a crowd. Not wanting to come across as a groupie, I avoided asking him all the usual questions, especially since I already knew his story so well. Instead, we talked about his life after retiring from NASA and he asked me about Canada's space program. He was not only an exceptional astronaut who had risen to the occasion when the stakes could not have been higher; he was also a man of surprising modesty and humility, qualities I greatly admire. I have rarely been so inspired by someone, and it was a great privilege to spend some time with him.

In early 1999, as Pam started training for another marathon, she announced that she was pregnant again, with a delivery expected in September. At this point, you'll forgive me for thinking there was a connection between running marathons and getting pregnant.

Fortunately, we were ready. Pam's pregnancy went smoothly, and to maintain tradition, we dined at Frenchie's the night before the delivery, again ordering our favourite pasta pappardelle. This time, a caesarean was scheduled, and I brought my camera. Everything went well, and on September 8, George Xavier Garneau joined us, weighing in at ten and a half pounds, quite remarkable considering that Pam is not a big woman. I was able to take some memorable "first" photos of him as he was lifted out of his mother's womb.

As the development of the ISS progressed, several astronauts, including myself, were participating in evaluating the ThinkPad laptop software and procedures that would allow ISS crews to interface with the various station systems. In parallel, NASA and the ISS partners were finalizing the sequence of assembly flights.

As always, the Astronaut Office had to choose mission crews with the appropriate combination of skills and experience. It was clear that assembling the station would require lots of spacewalking and robotics, not to mention rendezvous operations to dock with ISS. These were complex missions. As an astronaut with robotic experience, I was hoping for a mission where I could once again operate the Canadarm.

In late 1999, I received "the call" assigning me to mission STS-97 (also known as ISS mission 4A, reflecting its order in the assembly sequence). Because the flight was "weight-critical," the crew would be limited to five people. The mission commander would be Brent Jett, and the pilot, Mike Bloomfield. Mission Specialists Joe Tanner, Carlos Noriega, and I would make up the rest of the crew.

We would dock with the station and deliver a large truss segment known as P6, which carried the first U.S. solar arrays to power the ISS. When fully deployed, these would extend almost 240 feet, the largest arrays ever flown in space. Mike and I would both operate the Canadarm, and I would also have a number of other responsibilities, given our small crew size.

I was called by the head of the Astronaut Office shortly after being assigned to STS-97 and asked if I would accept being transferred to another flight, where I would also operate the Canadarm. The flight in question was a Hubble telescope servicing mission involving a great deal of robotic operations. I was given the option of saying no, and although the mission certainly interested me, given the importance of the Hubble, I asked to remain with STS-97.

Now that our crew had been assigned, it was necessary for us to travel to Russia to familiarize ourselves with their segments of the ISS. What struck me when we arrived in Star City was the stark contrast with the Johnson Space Center. The buildings had all the essentials but were drab in appearance, like so much Soviet-era architecture, and appeared poorly maintained. The halls were badly lit and some electrical outlets in the classrooms did not work. Although you could not argue with the success of the Russians' space program, it was clear that they operated on a shoestring.

The fact that Russia was a partner in the International Space Station was in many ways remarkable, considering the decades of Cold War that had preceded the breakup of the Soviet Union. I remember standing in the middle of Red Square with some American astronauts. One of them (not from my crew), who had previously been an SR-71 "Blackbird" high-altitude reconnaissance spy plane pilot, said that he found it hard to believe he was standing in Red Square, regarded as ground zero in the days of the Cold War, and now working with the people who had been his sworn enemies. Based on my own previous experience as a naval officer, I could certainly relate. Fortunately, the Cold War stayed cold, and we had taken a step back from the precipice.

I also had the opportunity to visit the Baikonur Cosmodrome site in Kazakhstan. This was nothing less than a pilgrimage for me, a priceless opportunity to see the birthplace of human spaceflight, the place from which Yuri Gagarin had been launched into space. I visited the small house where Gagarin slept the night before his historic flight. His bed was still there. I also witnessed the successful launch of a Soyuz rocket, the workhorse of the Russian space program. Touted as the safest rocket in the world, with over 1,900 flights, the first Soyuz rocket was launched in 1966, and despite undergoing changes over time, it remains the Russian vehicle for human space-flight. Canadian astronauts would eventually lift-off from Baikonur, beginning in May 2009, when Robert Thirsk flew to the ISS as part of the Expedition 20 crew, returning in a Soyuz re-entry vehicle six months later, landing on the steppes of Kazakhstan. Chris Hadfield and David Saint-Jacques would follow in his footsteps. For many years, the Soyuz rocket would be the only way to get to the station after the space shuttle was retired in 2011.

This was not my first visit to Russia. I had gone to Moscow in 1987 to represent Canada at celebrations commemorating the thirtieth anniversary of the launch of Sputnik, the first spacecraft to orbit Earth. In 1987 it was still the Soviet Union, a different place. I was assigned a room at the Rossiya Hotel just off Red Square, where most foreigners had to stay. On arrival, my passport was taken from me, and I was told it would be returned on my departure. This made me a little uncomfortable, and I confess to checking my hotel room for hidden microphones and cameras. I was still a serving naval officer at the time. On the positive side, I met some of the pioneers of the Soviet space program, among them Valentina Tereshkova, the first woman in space, and Alexei Leonov, the first human to do a spacewalk. Russians were clearly proud—and rightly so—of being the first in space on many fronts.

I would spend most of 2000 getting ready for my mission. I was now fifty-one and had decided that this would be my last flight. I had two young sons and felt I owed it to Pam to transition to a more "normal" life. I loved being an astronaut, but I also knew it wasn't always easy for my family, especially around launch time. Despite this, they had allowed me to indulge my passion for seventeen years, without ever asking me to give it up.

When I made that decision, I had no idea what I would do next. My plan was to start giving it real thought after my flight. But I had previously indicated to Mac Evans, the president of the Canadian Space Agency, that I intended to return to Canada. He contacted me in the spring of 2000 and offered me the job of executive vice president of the agency when I returned. I accepted. I wanted to remain involved with space and believed the position would offer me an interesting new challenge, as well as an opportunity to live in Montreal, a city that Pam and I both liked.

On ISS mission 4A, my main responsibility with the Canadarm was to lift the P6 truss out of the payload bay and position it close enough to the Station's Z1 truss that it could be secured by my crewmates Joe Tanner and Carlos Noriega, already positioned nearby, tools at the ready. The 15.8 metric tonnes P6 truss was 18.3 metres long, occupying most of the payload bay. The tolerances for extracting it were tight. Once I lifted it out of the bay, I would move it to an overnight park position in order to warm its components, and the next morning move it to where it could be secured to the station, with my two space-walking crewmates verbally guiding me as I moved it the last few feet. To assist with this task, I would also use another piece of Canadian technology, the Orbiter Space Vision System, which enabled a precise determination of the P6's position and orientation as it was being moved by the Canadarm to its connection point. This equipment had been designed by the NRC and built by Neptec, a Canadian company based in Kanata, Ontario.

Because we were a small crew, being able to replace each other on certain critical tasks was important. For that reason, I received EVA training as a backup to Joe and Carlos, who would be performing two spacewalks, including the one to connect the P6 truss. If for some reason one of them could not perform his spacewalk, I would replace him, and pilot Mike Bloomfield, who was also training on the Canadarm, would perform my robotic operations. This required me to undertake a smaller subset of Joe and Carlos's training in the Neutral Buoyancy Lab (the large swimming pool). Typically, EVA training sessions lasted about six hours and were physically demanding, especially when working upside down. The advantage of doing them was that I would better appreciate what my two colleagues had to do. This was important because I would be supporting them from the flight deck while they performed those tasks. I would be guiding them through the timeline, with an eye on the clock, and answering any questions they might have. That said, Joe and Carlos were so proficient that they didn't need any assistance. They knew exactly what to do and when to do it.

I was also designated mission specialist 2, the crew member who sits behind the commander and the pilot during launch and re-entry to provide any required support. My job was to watch the displays and status lights for fault indications and be ready to step up if a malfunction occurred. I would be holding the procedures manual on my lap and could confirm to the commander and pilot the steps to follow to address the malfunction. In preparation for this, I spent a great deal of time in simulators, learning to recognize various malfunctions and how to respond to them accordingly. In addition, I would be supporting the commander and pilot when *Endeavour* approached the station to dock with it, and later when it separated from it.

I was also designated as the crew "doctor" in case medical assistance was required. I was trained to perform various procedures such as drawing blood, doing an intubation and a tracheotomy, stitching skin, freezing a nerve before a tooth extraction, and dispensing medication. I really enjoyed this part of my training, which allowed me to be a "pretend doctor." There were even volunteer JSC employees who let me draw blood from them! That said, it was reassuring to know that a flight physician in Mission Control could guide me every step of the way should my services be required.

Finally, I would be responsible for transfers between the space shuttle and the station. We would be taking food, water, and new experiments to the ISS crew and bringing back waste and completed experiments. Supplying the crew was important for obvious reasons, but it was also necessary to bring waste and equipment back to Earth to ensure the station didn't become too cluttered.

This mission was different from my previous flights in two major ways. For the first time, I would be seated on the flight deck in my role as mission specialist 2. And secondly, we would rendezvous with another crew already on orbit. I could hardly wait for the ISS to come into view as we approached it to dock.

As planned, we launched on November 30, 2000, on a mission that would last almost eleven days. This was my second flight aboard *Endeavour*, and the first time I launched at night. As usual, I was secured in my seat about two and a half hours before liftoff. As the countdown proceeded, I went over the procedures we would use if something went wrong, procedures I had to have at my fingertips. And yes, as usual, I felt those butterflies.

Endeavour lifted off on time, rising majestically into the night sky. As with my previous flights, going to space would offer me another priceless opportunity to look down at planet Earth and marvel at the mystery of creation. I would savour it to the maximum, knowing this was my last flight.

Once we reached orbit, our first task was to rendezvous and dock with the space station, already occupied by one American and two Russians: Expedition Commander Bill Shepherd, Soyuz Commander Yuri Gidzenko, and Russian Flight Engineer Sergei Krikalev. They were known as the Expedition One crew, the first crew to live and work on the station, having arrived about a month before we docked to the ISS. We would be their first visitors, although we would not meet them face-to-face until we had connected the P6 truss, extended its two large solar arrays (port and starboard), and performed a number of other EVA tasks.

Robotic operations to extract P6 from the payload bay and position it for latching and bolting to the station's Z1 truss went off without a hitch, but the deployment of the starboard solar array uncovered a problem. Each array consisted of solar panels folded accordion-style inside a box. Deploying an array to its full length involved activating a motor-driven extension mast that slowly pulled the folded panels out of their box, unfolding them until they were fully extended. Unfortunately, some of the panels in the starboard box were stuck together, resulting in a jerky motion when they became unstuck and unfolded suddenly during their extension. When the array was fully deployed, we found that the two tension cables that were part of the extension mechanism had come off their tension reels, meaning the array could no longer be retracted, as would be required on a later mission.

The problem with the starboard array extension had not been anticipated and required the crew and Mission Control to devise a different procedure for the port solar array. Instead of extracting it completely without interruption, we paused the extension briefly every eight seconds, so that less force was imparted to the extension mechanism and tension cables whenever panels became unstuck. This modified procedure was successful.

Once both arrays were fully extended, an unscheduled third spacewalk had to be planned, designed, and executed to fix the problem with the tension cables on the starboard array. This was NASA at its finest, as the crew and a special team of experts on the ground developed the procedures for the additional spacewalk to fix the problem. And fix it they did. Although Carlos and Joe had not trained beforehand for this specific task, they executed it flawlessly, a clear demonstration of their spacewalking skills.

Almost a week after docking, all of us finally met face-to-face: five Americans, two Russians, and one Canadian. Hugs were exchanged and we shared a meal, which included Russian food brought up by Sergei and Yuri. We also shot some IMAX footage, visited the Russian and American parts of the station, and invited the ISS crew to the orbiter flight deck to enjoy the spectacular view of the solar arrays. We then began transferring the waste, equipment, food, and water. The latter included large water bags I had filled in the preceding days. The pure water was a by-product of *Endeavour*'s hydrogen fuel cells, which generated the orbiter's electrical power.

After little more than twenty-four hours together, it was time to say goodbye. With all transfers completed, hatches were closed and *Endeavour* undocked. Mike Bloomfield then flew us in an arc around the station while Joe and I photographed it and sent the pictures to the ground. It was a spectacular sight to behold. We had added an important element to the station, the first of four solar wings. Now it was time to return home.

As I savoured my last moments in space, knowing I would not be returning, I was conscious of the fact that life had handed me an extraordinary and rare opportunity. It had been a transformative experience, changing the way I viewed life on Earth, something I had certainly not foreseen at the outset. It had also been a privilege to share those moments with Canadians. In some ways, my astronaut career had been like a dream, but a dream that was real and that lasted seventeen years, allowing me to do what few of us ever imagine.

Was there a sense of regret that I would never again experience the exhilaration of spaceflight? While the desire never really goes away, the answer is no: I had no regrets. After three flights, it was time to make way for others. It was time for me to do something else.

When you're an astronaut, you strive to do the best you can. Looking back at my career, including the three missions I flew and the many years I worked in various support roles on the ground, I felt proud of what I had accomplished. That said, I cannot claim perfection. I made two mistakes on my last flight. One occurred in preparation for a spacewalk when I assisted crewmate Joe Tanner to put on one of his gloves by snapping the metal ring in the glove into the ring in the sleeve of his suit upper torso. After connecting it, I failed to engage a safety lock, to ensure it could not accidentally detach, which would result in a loss of suit pressure. I must have become distracted and forgotten this step. Fortunately, it was caught by Commander Brent Jett when he was doing final suit checks. The other mistake occurred after the first EVA, during spacesuit battery recharging in the airlock, when I disconnected the multi-pin electrical plug used for charging before it was unpowered. Removing power before disconnecting was a safety measure to avoid the possibility of shorting between pins. Fortunately, this did not cause a problem, but it was a mistake, nonetheless. All this is to say that, even with the best training and preparation, mistakes are sometimes made, and my performance was not flawless.

I described these two mistakes to the Astronaut Office during the STS-97 postflight debrief. This is something important about the culture that exists in the astronaut corps and more broadly within NASA. While NASA is widely admired because it often accomplishes the near impossible, it's important to remember that it has had to deal with technical challenges that sometimes led to tragedy. Sadly, we saw this with the *Challenger* disaster and later with *Columbia*. In both cases, hardware failed because potential failure scenarios were not fully understood or addressed.

There is always the starting assumption when designing an aircraft or space vehicle that there is no room for error, and this premise requires a particular approach—an approach based on total transparency. But beyond the rigorous design and testing that takes place at the beginning, there must also be a willingness to recognize problems that may come to light only at a later time, after a vehicle is already certified as operational. It means admitting that an oversight occurred, that a mistake was made. This is sometimes a greater challenge.

In the space business, owning your mistakes is more than important— it's critical. It is deeply ingrained in the astronaut culture and one of the reasons astronauts achieve excellence: rather than dwell on their successes, they focus on learning from their mistakes. That requires complete honesty. In that manner, everyone benefits, and the same mistakes can be avoided by others. Having been a capcom on many flights, I was present for every training debrief that followed a session in the simulator. The purpose of the debrief was to go over what went right, but even more importantly, what went wrong, either with the crew or in Mission Control. What struck me, whenever a mistake was made, was the willingness of individuals to say: "That was my fault." Honesty ultimately enables excellence. Looking for excuses has the opposite result. It should be in the culture of every organization.

Our re-entry went as planned and we landed at Kennedy Space Center at 6:04 p.m. on December 11. Because of my position on the flight deck, allowing me to see through the forward and overhead windows, I witnessed a spectacular lightshow and dramatic high-angle orbiter rolls as *Endeavour* ploughed back into the atmosphere, bleeding off speed (and energy) as it slowed from Mach 25 to two hundred knots at the threshold of KSC's runway. For about 10 minutes we were enveloped in an orange glowing plasma, visible through the forward cockpit windows, with light flashes streaming past the overhead windows as the orbiter compressed an increasingly thickening atmosphere, causing it to heat up rapidly. Thank God for those protective tiles, blankets, and carbon coatings! Unquestionably, an unforgettable and dramatic way to bring my astronaut career to an end.

Let me mention one other duty I considered a privilege during my time in Houston. I was a family escort on three occasions. When a crew is named to a flight, two astronauts are chosen to assist the crew's families during the time the crew is gone. This becomes critically important if something goes wrong. Family support begins the moment the crew goes into quarantine and ends when they return. There are well-established procedures to support families if the worst should happen. Going over these procedures before a flight drives home just how much is at stake for your own family when you fly. Fortunately, everything went well on the three occasions I was a family escort. Sadly, this was not the case when *Challenger* and *Columbia* were lost.

The last flight for which I acted as a family escort was STS-93. This flight was significant because its commander was Eileen Collins, the first woman to command a space shuttle mission. Given the historical nature of the flight, First Lady Hillary Clinton and her daughter, Chelsea, attended the launch, which occurred on July 23, 1999. There are elaborate protocols to follow anytime the president or the first lady attends a NASA event, and these required an astronaut to be available to host Mrs. Clinton and her daughter for about an hour before the launch to answer any questions they might have. Although several American astronauts were available, the task was assigned to me. The first lady and her daughter were surprisingly well informed about what was about to happen, and the launch went well, so my job was easy.

Shortly after *Endeavour* landed, Pam and I started packing for our return to Canada. It was also time to say goodbye to our friends and neighbours. The movers arrived a few days after Christmas and we watched, not without sadness, as the house in which we had started a new family and lived for more than eight years was emptied.

This was it—the end of my astronaut career. I remembered how I had felt when I retired from the navy, my first love, and this was really no different. Both careers had been immensely satisfying, but now it was time for me to do something else. It has always been important for me to keep moving forward, and my mind was already transitioning back to where I came from. With two young children in tow, Pam and I were eager to settle in Montreal, and I was excited to be taking on my next challenge—helping to shape Canada's future in space.

EIGHT

GETTING USED TO THE CANADIAN WINTER after eight and a half years in Texas was surprisingly easy. Although they had never experienced it, Adrien and George (and four-legged Rosie the Bouvier!) took to the snow like ducks to water. It felt good to be back in Canada. We were ready to start our new life. I felt I was where I should be.

My place of employment, the Canadian Space Agency, had its offices on the edge of Montréal/Saint-Hubert Airport, on the South Shore of the St. Lawrence. The structure itself was attractive and quite original in its design, most unusual for a federal building. Seen from the air, it somewhat resembled a large spaceship, with its curved front and the three building wings extending behind it. There had been a major political tug-of-war about its location (Ottawa versus Montreal), and in the end, the decision was made to build on Montreal's South Shore.

As I began my new job as executive vice president, I sought inspiration from many quarters, conscious that I had been given a big responsibility. The CSA was an agency of the federal government, in service to Canadians—but what exactly did that mean? Perhaps no one said it better than John H. Chapman back in 1967, when he was leading Canada's efforts to use communication satellites to link Canadians across our vast expanse: "In the second century of Confederation, the fabric of Canadian society will be held together by strands in space just as strongly as railway and telegraphy held together the scattered provinces in the last century." It was a compelling vision. What I took from it was that using space would enable the CSA to dramatically improve the lives of Canadians. It would be my job to make sure we did that.

Although Canada had been involved in space since the late 1950s, its various programs had been managed separately by different government departments. It was not until 1989 that a decision was made to consolidate those programs within one organization, the CSA. Now, for the most part, programs dealing with Earth observation, satellite communications, space science, and human exploration would be managed by the CSA. (The exceptions were astronomy and astrophysics, which would remain with the NRC.)

At its core, the mission of the CSA was straightforward: to develop space programs that would benefit Canadians, further science, and foster a competitive space industry. To execute its mandate, the CSA employed professionals from a range of fields, including engineers, scientists, and program managers. In addition to its own national programs, it would join international initiatives with other space agencies. The CSA reported to the minister of industry, science, and technology and received an annual budget of about $300 million, with additional funding for certain major programs.

When Mac Evans had offered me the position of executive VP, it was to prepare me to replace him as president sometime in the fall of 2001. I had no doubt I would have big shoes to fill. Mac had been nothing short of a pioneer in Canada's space program, beginning with his role as mission director for the Communications Technology Satellite program, which culminated in the launch in 1976 of the Hermes spacecraft in partnership with NASA. At the time, Hermes was the most powerful communications satellite in the world. Later, Mac would lead negotiations for Canada's participation in the International Space Station program. He also oversaw the development of Canada's first space plan, which resulted in the Canadian Space

Agency. Throughout this period, he played key roles in Canada's RADARSAT and astronaut programs. For all these reasons, I considered him a mentor.

As I took up my duties, I realized I needed to both broaden and deepen my knowledge of Canada's space program. While I was familiar with its main elements, my knowledge in certain areas was superficial and I would have to dig much deeper. I needed to learn more about satellite communications and Earth observation, two domains of critical importance to Canada. I also needed to learn about our accomplishments in space science. It was, after all, for scientific reasons that we first went to space with *Alouette 1*, a spacecraft that Canadians, under the leadership of Dr. Colin Franklin, designed and built to study the ionosphere, an electrically charged layer of the Earth's upper atmosphere. Along with human exploration, these were the main elements of our space program and areas in which Canada already had a well-established track record. I was excited to be working with a group of highly motivated professionals and eager to learn from them. One of many of these was Dr. Virendra Jha, who had immigrated from India as a young man and played a key role in Canada's space program, first in the private sector at Spar Aerospace, eventually becoming its director of engineering, and later as CSA's vice president responsible for science, technology, and space projects.

In the late summer, as anticipated, I received a call from Percy Downe, the director of appointments in the Prime Minister's Office, asking me on behalf of Jean Chrétien to assume the presidency of the CSA, an assignment I accepted with pleasure. After eight months of getting up to speed, I felt I was ready for the job. On September 28, Brian Tobin, the minister of industry, announced that I would take over from Mac Evans on November 22, 2001. On that date, I assumed full responsibility for Canada's space program.

As we adjusted to our new life in Canada, Pam and I began to participate in the cultural life of Montreal. The opportunity arose when I received a call from Boris Brott, the musical director of the McGill Chamber Orchestra. I had known Boris for many years. We had done a dozen concerts together, where he conducted classical music and, between pieces, I would talk about being an astronaut and answer questions he would ask me. He chose music with a space theme, such as *The Planets* by Gustav Holst, or music I had taken with me on my first flight. As he conducted, slides of various planets, galaxies, and constellations would appear on a large screen behind the orchestra.

Boris had approached me with the idea for these concerts after my first flight in 1984, after he heard me speak on the radio about listening to classical music in space. I liked his idea and we ended up performing our concerts throughout that decade in Canadian cities and even once in California. They were quite popular with the public and particularly with children. For the record, I do not play any musical instrument but have always been impressed by those who do. My love of classical music came from listening to it in my home, beginning at a young age, when my father introduced me to Tchaikovsky's *1812 Overture*.

Boris now asked me to become president of the McGill Chamber Orchestra, with the objective of moving its finances from the red into the black. I agreed to his request and would do the job for the next four years. The challenge ahead of me had nothing to do with quality. The McGill Chamber Orchestra (later renamed Orchestre classique de Montréal) had been founded in 1939 by Boris's father, composer and conductor Alexander Brott, and his mother, cellist Lotte Brott, and had a well-established reputation. The challenge was that Montrealers had so much to choose from, with two full-sized symphony orchestras and several chamber music ensembles, making the competition fierce to attract a finite audience. Gradually, after hiring a solid manager, Susan Bell, and bringing in new board members who knew how to fundraise, we were able to turn things around and make the orchestra solvent. That was my small contribution to the artistic life of Montreal, something I am quite proud of.

A sad footnote: as I was writing this chapter, I learned of Boris's tragic death at the age of seventy-eight after a hit-and-run while he was out for a walk in Hamilton, where he lived—an unspeakable loss for the Brott family and for Canada, which lost one of its most distinguished citizens.

Over time, Pam and I made new friends and our children adapted well to their new surroundings. In the winter, the City of Westmount erected an outdoor ice rink at the corner of our street. Adrien, who had started with ball hockey in a neighbour's driveway in Houston, would be out there for hours, by himself or with friends, learning to play on ice. Some nights, it was like pulling teeth to get him home for dinner. He and George would eventually play in the Westmount hockey league, with Pam and me cheering them on from the bleachers. Meanwhile, Rosie was also thriving in Canada's winter after the oppressive heat of Texas. She would often plunge her head deep into the snow, seemingly just for the pleasure of it.

That summer, Pam and I began our tradition of renting a cottage in the Laurentians to get a break from the city. I loved being in the woods, on a quiet lake, watching the sun set and listening to the loons, a quintessentially Canadian experience that always put my life back in perspective.

At work I continued to broaden my knowledge of Canada's space program, and the more I learned, the more I realized the proud legacy of our space pioneers, in science with *Alouette 1*, or in communications with the Hermes satellite, in Earth observation with RADARSAT-1, or in space robotics with the Canadarm. It was up to me to build on that impressive legacy.

Canada had gone to space for practical reasons. Building *Alouette 1* to study ionospheric physics not only made Canada a world leader in this domain; it also helped us better understand how auroral—or geomagnetic—activity affected high-frequency communications over Canada's North, an issue of practical importance at the time because, until the advent of communication satellites in geosynchronous or low Earth orbit, high-frequency radio waves bounced off the ionosphere were used to communicate over long distances.

Operating communication satellites allowed us to connect to Canadians living in remote regions of our country. Designing a spacecraft like RADARSAT-1 enabled us to monitor our lands, lakes, rivers, and oceans, day or night, regardless of cloud cover, a requirement that was becoming increasingly important with climate change. In other words, there were compelling reasons for Canada to be in space, and that need will only grow in the coming decades. Succinctly put, Canada and space were a natural fit, and we had a proven track record of excellence in this challenging domain.

While the public is often most interested in human spaceflight, going to space is about much more than humans living on a space station or going to the Moon or eventually to Mars. It's also about building the invisible infrastructure that orbits high above us and helps us monitor what is happening to our planet, communicate with each other, predict the weather, measure our precise location anywhere on Earth, and understand the origins of our solar system and of the universe.

I had two objectives in mind when I took the helm of the CSA: to produce a coherent space strategy that would take Canada into the future, and to make Canadians realize how important space was in their lives.

My strategy focused on our three primary stakeholders: the government, the industrial sector, and the scientific community. For government, space was a valuable asset to meet objectives related to security and sovereignty, connecting Canadians, and monitoring the overall health of our country. For industry, space offered the opportunity for our manufacturing and service sectors to grow, based on their well-established track record. For the scientific community, space offered the potential for new discoveries.

That being said, I was beginning to think that we should be doing more. Why? Because we needed to be in space and because we were good at it. There were more initiatives that we could take on, particularly in Earth observation and understanding the causes and effects of climate change, matters of some urgency. But without additional funding, we had to limit our ambitions.

We had been the first country in the world to control our own communication satellites (the Anik series) when the government created Telesat Canada in 1969; we pioneered the use of Ku-band and later Ka-band transponder frequencies in satellite communications, providing greater capacity and direct-to-home service; we made important advances in Earth observation using synthetic aperture radar with RADARSAT-1, which produced images that are sold around the world; and we designed and built the space robotics for the space shuttle and the ISS.

Canadian space companies had a well-deserved reputation for delivering the goods, while Canadian scientists and astronomers were praised for their world-class contributions in advancing our knowledge of planet Earth, our solar system, and indeed the universe.

As an example, I must mention the James Webb Space Telescope, now delivering spectacular images from its observation point 1.5 million kilometres from Earth. Canada is an active participant in this ambitious project, having contributed the Fine Guidance Sensor, which enables the telescope to point at and focus on particular objects, and the Near-Infrared Imager and Slitless Spectrograph, which helps us study many astronomical objects. Webb is only one example of the many space science projects in which Canada and the U.S. have cooperated over many decades.

Speaking of telescopes, not all of them need to be launched into space. A great deal of astronomy can be done from the ground, depending on what wavelengths are being captured. This is far less expensive than building space-based telescopes and allows direct human access for observations as well as for equipment maintenance and upgrades. Certain locations on Earth are better than others because of the clarity of the sky above them, where the atmosphere has minimal pollution, water vapour, or cloud cover and, equally important in some cases, minimal stray lighting. Places like the Atacama Desert in Chile or the summit of Mauna Kea in Hawaii (home of the Canada-France-Hawaii Telescope) come to mind. That said, astronomy at certain wavelengths of the electromagnetic spectrum can only be done with space-based telescopes such as the Hubble and other telescopes that must be located above Earth's atmosphere to clearly capture those wavelengths.

During my time in Houston, I had come to appreciate the sheer breadth of the American space program. At the time, NASA's budget was roughly sixty times Canada's, and that didn't include the U.S.'s military space program, a major recipient of federal funding. For a country whose economy was roughly eleven times smaller than that of the United States (based on 2021 GDPs), you had to wonder whether Canada was investing enough in space.

During my four-year tenure as president, the annual budget of the CSA moved very little. While the Liberal government, under Prime Ministers Jean Chrétien and Paul Martin, was proud of our space accomplishments, the predominant view was that the CSA was adequately funded. Decision-makers saw that astronauts were flying, that we were contributing robotics to the ISS, that the RADARSAT-3 constellation was in the preliminary design phase, and that other, smaller projects were underway. What they had trouble seeing was Canada's unrealized potential in space.

If Canada's space budget was to remain modest, we had to find imaginative ways to leverage our presence in space. The best way was to work with other countries, as we had done with NASA since the early 1960s. That's why Canada signed an agreement with the European Space Agency in the 1970s, as well as several bilateral agreements with other countries that allowed us to fly Canadian hardware on their spacecraft in return for sharing the scientific results. These bilateral arrangements also provided useful exposure for Canada's space industry. Maintaining those relationships will continue to be important.

In 2005, I travelled to China to meet with Sun Laiyan, head of the China National Space Administration, to explore the possibility of space cooperation in a "non-sensitive" area such as upper-atmospheric physics. China had embarked on an ambitious space program and there was no doubting its resolve. It was making space a priority, with the aim of catching up to the United States. Without making too fine a point of it, the China of 2005 was not the China of today.

The trip allowed me to see just how much their space program had progressed. They were proud of their Long March rockets, their spacecraft launches, and their new human spaceflight program. During our stay, we were given a tour of a test facility on the outskirts of Beijing and met with various officials. This was preceded by a formal meeting between Sun Laiyan and me, complete with interpreters. In his remarks, Sun mentioned that he had studied in Paris, which prompted me to say a few words in French, at which point his eyes lit up. We continued in French and the conversation became more informal. Though the trip was successful and a follow-on meeting was discussed, nothing further happened after Stephen Harper became the prime minister the following year. Needless to say, relations have become more challenging in recent years, and cooperation in space at this time is not on the table.

In 2002, Queen Elizabeth celebrated her fiftieth year on the throne, her Golden Jubilee. To mark the occasion, in October, Governor General Adrienne Clarkson invited fifty Canadians to Rideau Hall for a luncheon with Her Majesty, who was visiting Canada. Each invited Canadian represented one of the fifty years of her reign, and my year was 1984, the year of my first spaceflight. Pamela was with me, and we had the pleasure of joining many Canadians we had both long admired: author Michael Ondaatje, film director Atom Egoyan, Stompin' Tom Connors, Doris Anderson, Margaret Atwood, Oscar Peterson, and many others. We also had the opportunity to meet the Queen and Prince Philip and to chat briefly. I was introduced as a Canadian astronaut, and this prompted the Queen to ask me questions about space. I have to say that I was taken with her, even though I'm not a monarchist. I could see why she was a beloved figure.

An amusing incident occurred that day between the Queen and Louis Garneau, one of the fifty guests. Louis (no relation, unless you go back very far) was a well-known Canadian cyclist turned businessman. He wanted his picture taken with the Queen and asked his wife to take it as he boldly sidled up to the monarch and put his arm around her with his hand resting on her shoulder. This provoked gasps from the British entourage accompanying the Queen, but the Queen herself did not seem upset in the least and gave a big smile. I think that was one of the reasons people liked her so much.

In late 2002, I was invited by Carleton University to become its chancellor. The original plan had been for Ray Hnatyshyn, our previous governor general, to occupy the post, but due to a sudden and aggressive cancer, he had to bow out at the last minute. Sadly, pancreatic cancer would take his life in a matter of months.

I was honoured to be offered this role, a responsibility I would fulfil for five years. My job was to confer degrees on all graduating students and to briefly greet those who chose to attend their convocation. I enjoyed observing the expressions on their faces as they crossed the stage and we saluted each other, sometimes with a handshake, sometimes with a nod of the head. Some were nervous while others swaggered across the stage. I shook dry hands and I shook moist hands. Some were dressed to the nines, others were wearing shorts and sandals, looking like they were on their way to the beach. Regardless, every student looked profoundly relieved to have reached this moment, and deservedly so. It reminded me of my own graduation. It was a major milestone for each of them and concrete proof that they had accomplished something important. In some cases, the joy on the faces of their cheering relatives said it all.

On a personal level, I found the role of chancellor to be tremendously fulfilling, and I never took it for granted. It connected me to students celebrating an important achievement in their lives. Such an experience rubs off on you in a positive way. You are witnessing progress, which is energizing. In the years following my tenure, many graduates have come up to me to recall that moment in their lives.

In early 2003, we lost Rosie to cancer. She was only eight years old. We knew that bigger dogs did not live as long as smaller ones, but losing her still came as a shock. We tried everything, including chemotherapy, but when that failed, we decided we did not want her to suffer any longer. One quiet Sunday morning, we called our vet, Allan Gilmour, who kindly came to the house. We all gathered around Rosie, who was lying on her side in the corner of the living room, her feet tucked under the carpet, as was her habit, and we said our goodbyes to her before Allan put her to sleep. Slowly, she closed her eyes and stopped breathing. It was a peaceful but difficult moment for all of us, including Adrien, who was seven, and George, who was four. I was surprised at my own level of grief. I had not grown up with dogs, so this was new for me. We all loved Rosie and felt the void left by her departure. We had so many fond memories of her growing up in Texas and her life with us in Montreal.

She had loved running in a particular field near our home in Clear Lake, and so we decided to spread her ashes in that field. We put a portion of them in a container that I took to Houston on my next visit. I still remember being stopped by security at Montreal airport and the official removing from my carry-on bag a Tupperware container marked "Rosie's ashes." He looked at the container and then at me, and with a sad expression, offered his condolences for my loss. No doubt he thought that Rosie was a human relative, but I was still touched by his compassion. That evening, I walked over to Rosie's field, which was deserted. I lifted handfuls of her ashes and threw them in the air around me until they were all gone. Later that year, we brought another puppy into our lives, a goldendoodle we named Lola. She would be with us for seventeen years.

—

Those who work in the space business know that everything we undertake is high risk, not just human spaceflight. Success is achieved only if everything goes perfectly. It's always leading-edge technology, costs a lot of money, has to withstand a violent ride to space, and must operate for years in a hostile environment. There is no way to gently put a spacecraft in space and retrieve it if it doesn't work.

That was driven home for me during the development of the RADARSAT-2 spacecraft, the next-generation version of the successful and long-lived RADARSAT-1. This type of spacecraft observes Earth with a sensor known as a synthetic aperture radar. Canada is a leader in this technology, processing the radar data back on Earth and transforming it into useful images sold commercially around the world, particularly in locations experiencing natural disasters. Of particular importance to Canada are images showing ice cover in the Arctic, on our Great Lakes, and off our coasts.

The cost of RADARSAT-2 was roughly twice the annual budget of the CSA, and consequently, a great deal was at stake when it was launched in 2007. Any number of things could go wrong: the rocket carrying it could explode or fail to reach its proper orbit, or the spacecraft itself might not operate properly once it did reach orbit. Think of the money spent and the years of effort involved every time a launch occurs. I remember sleepless nights before the launch and subsequent checkout of Canadian spacecraft or payloads and heaving a sigh of relief when they were finally declared to be operational. Space will always remain a high-stakes game, and Canada's space industry can be proud of its record of success.

Space projects are not only complex and expensive. They also take years to complete, often five to ten years or even longer. For example, Canada's participation in the International Space Station was well underway when I arrived, and it became my responsibility to continue the work begun by my predecessor. Similarly, I would initiate projects like the RADARSAT-3 constellation, which my successors would take over the finish line.

On June 30, 2003, a small Canadian space telescope called MOST, which stood for Microvariability and Oscillations of Stars, was successfully launched, and sometime later the lead scientist, Jaymie Matthews from UBC, visited the CSA to discuss how the telescope was performing. He dropped by my office and brought an unexpected guest with him, Justin Trudeau. We met for about a half an hour. Trudeau mostly listened but also asked a few questions about Canada's space program. I don't know if he came out of general interest or because he was examining possible future options for himself. As I recall, he was studying engineering at the time. I thought nothing more of it. Who would have guessed that just a few years later both of us would be elected to Parliament? Certainly not me.

As it turned out, this was not the first time I had met him. He reminded me, after we both entered politics, that I had visited his class when he was in elementary at the Rockcliffe Park Public School in Ottawa, one of hundreds of visits I have made over the years to speak to young people.

One of the benefits of returning to Canada in 2001 was that I could now regularly see my parents. While my mother was thriving, my father had been diagnosed with Parkinson's. Over time, I watched it progressively take control of his body and eventually his brain. Being a stoic, he never complained and soldiered on. I cherished those moments when I could still communicate with him, but it was hard to watch this once proud man succumb to a merciless disease. I had looked up to him my entire life. He had guided me in moments of personal difficulty. Over time, his condition worsened and he was diagnosed with Lewy body dementia.

He died in January 2004, weeks from his eighty-third birthday. I was with him at the Perley and Rideau Veterans' Health Centre in Ottawa when he drew his last breath. It had been a long and painful ordeal for both him and my mother, lasting many years and, in my opinion, too long. It was heartbreaking for all of us to watch him succumb to his dementia, robbing him of his dignity, something we should all be able to preserve until the end. In his last year, he could communicate with us only through his eyes, and sometimes when I looked into them, I thought he was telling me to get him out of his hellish physical and mental prison. I'll never know. (I would often think of him later, in my political life, when I co-chaired the Special Joint Committee on Medical Assistance in Dying.)

In the end, I chose to remember him as the man on whose big shoulders I sat when I was a little boy.

More than three years into my mandate, a vision for Canada's future in space had taken shape in my mind. As I described in previous chapters, my own spaceflights had played an important role in how I viewed the world. The experience of seeing Earth from above strengthened my conviction that we all share this planet and that if we destroy its atmosphere, its land, and its oceans, there is no option B. We live on a beautiful planet, but it is not a place of inexhaustible resources and unending resilience—a view that may not be obvious to some Canadians, given the vast, resource-rich country we live in.

From space, I had seen the damage we are inflicting. I had seen our great forests burning, our soil being washed away through flooding and deforestation, and our atmosphere being polluted by vast clouds of smog. Our assault on our planet has been relentless. We need to act globally to reverse this course, beginning with a better understanding of what is happening. Earth observation from space can help us with this, and Canada already has considerable expertise in this area. We need to make it our most important priority.

We also need to continue encouraging the privatization and commercialization of space activities. At the beginning, any government that had ambitions in space had to fully fund its programs. No private enterprise was willing to invest in such a costly, complex, and risky business. Fortunately, through the leadership of early Canadian visionaries, the Canadian government took on an enabling role.

Working with the private sector, it provided the majority of the initial funding to develop certain capabilities and then transferred them to the private sector—for example, when it created Telesat, now a successful global communication satellite operator, or when it funded RADARSAT-1, the Earth observation satellite that allowed the space company MDA to sell its images to the world. In both cases, the government helped get things going, with the intention of creating commercial opportunity.

Not surprisingly, I also believe that it is important for Canada to make a long-term commitment to the human exploration of space. While space can serve our purposes in many useful ways back here on Earth, it can also play a role in exciting us, in inspiring us, and in helping us dream—something we all need. That happens when countries build incredible scientific instruments like the James Webb Space Telescope, but it also happens when we send humans to space and they share their experience with us.

Our decision to join others to build and operate the International Space Station had been the right one, as was the decision in 2020 to participate in NASA's Lunar Gateway project as part of the ambitious Artemis program. We want to be there when humans return to the Moon. We want to be there when humans embark on the journey to Mars.

If I were to summarize my position on the future of Canada's space program, I would say we need to ramp it up. I have always believed that Canada should be doing more, not just because we need space to help us here on Earth, but because we're good at it. At a time when it's important to focus on areas of innovation where Canada has a proven track record, we are failing to exploit one of our most obvious strengths, and we risk being overtaken by others. I had made this point repeatedly over the years, both as president of the CSA and as a minister in the government.

———

I was loving my time at the CSA and was hoping to be renewed for a second five-year mandate. Life, however, has a way of surprising you. Late in the summer of 2005, I received an unexpected visit from two Liberals, Hélène Scherrer, principal secretary to Prime Minister Paul Martin, and Senator Dennis Dawson. Their visit came out of the blue, and I had no idea what they wanted to discuss. Because I had an open-door policy, I was used to receiving a variety of visitors. For example, Guy Laliberté, the founder of Cirque du Soleil, dropped in one day to ask me some questions about space as he began exploring the prospect of a trip to the International Space Station on a Soyuz rocket, a trip he would later undertake.

As it turned out, my two Liberal visitors were not, like Guy, interested in a trip to space; rather, they were testing the waters. They had been given the task of scouting for Quebecers who might be willing to run for the Liberal Party in the next federal election, which might be only months away. This type of recruiting is common among all political parties. First, they asked me whether I was a Liberal and, if so, would I be interested in running? I was taken aback, and while flattered, realized this would be a big decision, requiring serious reflection.

I had always followed politics, because I found it interesting and I wanted to keep abreast of government decisions that might affect me. But I had no political experience. Not only that, I loved my job at the CSA. In many ways, this was like that moment twenty-two years before when I spotted the "Astronaut Wanted" ad. Applying had carried the risk of failure, yet I had taken the chance. Now, for the second time, I was facing a similar decision between a job I had and loved and a job that might open up exciting new possibilities but carried a real risk of failure.

The invitation to enter politics was compelling for one important reason: I knew it was the politicians who ultimately made the decisions that determined the course of our country. I could propose ideas to them, but it was they who decided. This had become abundantly clear to me during the previous four years, as I reported to the minister of industry, science, and technology. The possibility of making important decisions that would shape Canada's future appealed to me.

I had not grown up dreaming of becoming an astronaut, but it had come to be, in part because I had dared to believe in myself. Was I ready to take another sharp turn in my life? Was I ready to give up my love affair with Canada's space program and throw myself into something as wildly unpredictable and risky as politics? As an outsider to the political realm, I had always thought, despite the attraction, that you'd have to be a little crazy to enter the political arena. Was I a little crazy? Was I willing, once again, to face the unknown?

NINE

I HAVE ALWAYS BEEN A LIBERAL BUT A DECIDEDLY BLUE ONE, leaning slightly to the right of centre. I believe governments should provide a social safety net for those less able to make it on their own. That's the hallmark of a compassionate society. At the same time, I believe in fiscal responsibility, meaning that governments must keep deficits under control and minimize the national debt for the sake of future generations. It's a balancing act that often requires painful decisions and tough medicine.

I admired Paul Martin for making tough decisions when our finances were out of control in 1993. At the time, the *Wall Street Journal* had mused about making Canada an "honorary member of the third world"! Coming from the private sector, Martin knew better than most the importance of balancing the books, and although it was going to hurt, belt tightening was necessary. As a private citizen, I wasn't happy with the measures he imposed, but I recognized that they were essential if we were to put our house in order. Over time, he delivered surpluses, allowing us to reduce the national debt.

Historically, the Liberal tent has been large enough to accommodate people like me, whose thinking counterbalanced that of left-leaning Liberals. In my opinion, that balance was essential and key to the party's historical dominance. But while I held Liberal values, was I ready to embark on the uncertain journey of political life? I had only a short time to decide. Over the following weeks, I discussed it with Pam, weighing the pros and cons.

This was not the first time I had been approached to become involved in politics. Back in the late 1980s, when I was still an astronaut, Paul Martin, then CEO of Canada Steamship Lines, invited me to lunch. Because of my navy background, he wanted to know my views on whether Canada should acquire nuclear-powered submarines. It was a topic of debate at the time, the government having just published a White Paper on national defence. I supported the idea for the following reason: I knew that other countries with nuclear submarines were moving freely through Canadian Arctic waters and I felt that Canada should have a similar capability in order to send a clear signal that we were serious about our sovereignty.

Even though we only touched on politics, it was clear to me that Martin was preparing his own entry into the political arena while trying to gauge whether I might have a similar interest. I was certainly intrigued, but the timing was not right for me, having become a single parent with two children who needed my attention. There was also the hope, at the back of my mind, of a second spaceflight. Although I didn't follow up, our encounter left a lasting impression on me.

Returning to the decision at hand, I knew that after more than twelve years in power, Liberal fortunes were ebbing, and the sponsorship scandal was causing the government a great deal of trouble in Quebec. In that respect, a wiser person would probably have wanted to sit out the coming election, judging it to be unwinnable. Rightly or wrongly, I did not give this much consideration and focused instead on whether I could make the transition to political life. By nature, I was not a political animal. As well, there was a gladiatorial quality to politics that I wasn't sure I liked. Would becoming a politician change me in unintended and, perhaps, undesirable ways?

I was fifty-six years old, quite old to enter politics. That being said, my health was good and I felt energetic. I also had name recognition and real-world experience to bring to the table from my careers in the navy and space, something I felt was of value, even though I knew some politicians had been elected in their twenties, with no professional experience whatsoever.

If elected, I would be the first Canadian astronaut to become a politician. I could not help but think of John Glenn, not just because he had become a successful U.S. senator but also because he was hardly the modern political type—he tended to be private, serious, practical, and anything but showy and extroverted. I admired those qualities, and liked to think there was room in the Canadian political sphere for people like that.

It was Pam's unconditional support for whatever decision I made that tipped the scales. She, of course, would be affected by such a change of career, but she also knew me better than I knew myself. In early November, I called the party to say I would run. At the time, the thinking was that an election would occur the following March, although it could come as early as December, given the unpopularity of the Liberals. The earlier date proved to be the right one, as details from the Gomery Inquiry into the sponsorship scandal emerged, prompting the opposition parties to band together to defeat the government in a vote of non-confidence.

I had to hit the ground running. A quick visit was arranged to meet the prime minister, who welcomed me to the team and wished me good luck. His Quebec lieutenant, Jean Lapierre, wanted me to run in Vaudreuil-Soulanges, just west of Montreal. I would be replacing Nick Discepola, who had held the seat until 2004, before losing to Meili Faille from the Bloc Québécois when the electoral boundaries were redrawn to her advantage. Although Nick wasn't happy about stepping down, he graciously offered to brief me on some of the local issues.

I introduced myself to the riding executive and we started getting ready for the campaign, which would last fifty-six days at the darkest and coldest time of the year. The party appointed Hervé Rivet as my campaign manager, and while he was scrambling to assemble a team to support me, I tried to bring myself up to speed on the election issues. From the outset, the odds did not look good. The Liberals had been in power for more than a dozen years and there was a strong hint of corruption owing to the sponsorship scandal. We could sense that Canadians were thinking it might be time for a change. The way I saw it, there were two possible outcomes: either I won and would begin my political life, or I lost and would be looking for a new job. Nevertheless, I had crossed the Rubicon. While I did feel some guilt for leaving the CSA so quickly, I took comfort in the thought that I was leaving it in good shape. I quickly said all my goodbyes and plunged headfirst into the unknown.

My campaign did not begin particularly auspiciously. I had been told to meet Prime Minister Martin's campaign bus at a motel in the riding where I was running. Together, we drove to Coteau-du-Lac, a small community on the St. Lawrence, where a rally was being held at the arena. Because the prime minister was present, the media were also there in large numbers. Like the veteran that he was, Paul Martin worked the crowd, introduced me, gave a speech, and answered questions from the media. I felt a bit like a bump on a log but realized this was how election campaigns unfolded.

The media did ask me some questions on the way out, which I did my best to answer, probably looking like a deer caught in the headlights. It didn't take long for me to realize that I needed to acquire a lot of new skills. While I was at ease answering questions about space, I was not fully prepared for the aggressive in-your-face approach that is more common in scrums with political reporters, particularly during elections, nor was I ready for the broad range of topics on the table, some of them rather delicate and requiring careful consideration. Let's put it this way: it is far more challenging to talk about the sponsorship scandal than it is about how I felt launching into space. What I learned that day was that I needed to prepare more rigorously in order to anticipate the questions I would be asked, and be in a position to answer them.

Nationally, the Liberal campaign was not going well, and it was particularly difficult in Quebec, something that was becoming painfully obvious with each passing day. Neither I nor my party could afford to make mistakes. Unfortunately, both of us made our fair share, although I'll chalk most of mine up to being a rookie.

My first faux pas illustrates the challenges I faced in politics. A journalist asked me whether I would leave Quebec if the province ever separated from Canada. I said yes, I would. It's the kind of thing one might say to a reporter if you didn't know any better, which I did not. Of course it led to a great deal of unflattering coverage in Quebec's French media. Unwittingly, I had been caught in a classic trap. My answer should have been: "It's never going to happen and I will fight with every fibre in my body to make sure it doesn't." The point is, I had a lot to learn, and I had precious little time in which to do so. Trial and error is not the best formula on the campaign trail, something I learned the hard way.

I had been billed as a "star candidate," which was a doubled-edged sword. Yes, the media and the voters knew who I was and, accordingly, gave me the attention they perhaps did not afford other, less well-known candidates on the hustings. On the other hand, the media, especially, often put a big target on the backs of those considered "stars," and it was a mistake on my part to not fully appreciate this dynamic, nor that the journalists who cover politics are decidedly different from those who covered me in my space career. Frankly, it was naive of me not to realize this. Not only was I an inexperienced politician, but my instinct throughout my life as a naval officer, engineer, and astronaut was to answer questions factually and head-on. I lacked experience in recognizing the subtext of a question—the "gotcha" trap, if you like—so I paid a price for my candour, honesty, and inexperience. Well-seasoned politicians learn how to avoid those pitfalls. They learn what to say, when to say it, and how to say it. Most importantly, they learn what not to say. This may be a sad aspect of the political culture, but as someone who wanted to be a politician, I needed to suck it up, find the right balance, and adapt where necessary.

I was faced with another challenge. Because I was running in a half-rural riding, agricultural matters were important to the constituents. Although I was given excellent briefings on the issues of concern, there was a credibility gap. Local producers knew I was a city boy from downtown Montreal, and although they were invariably respectful, they also decided I wasn't one of them. It's also possible that because of my previous job as an astronaut, I was viewed as someone who could not relate to "everyday" people.

I also ran into strong separatist headwinds. The Bloc would end up with an impressive fifty-one seats, and their local candidate, Meili Faille, was quite popular. We sat together through several Christmas lunches, and it was obvious she enjoyed considerable support. That being said, I felt I was doing well in some of the suburban parts of the riding, home to many Montrealers who had moved off-island because of the less expensive real estate.

"I like you as an astronaut but I won't be voting for you because you're a Liberal" was an all-too-familiar refrain for me. Less generously, many voters referred to the Liberals as "just a bunch of crooks." Most people were respectful when I interacted with them, regardless of whether they were going to vote for me. People who shouted at me, made jokes about me still being in space, painted swastikas on my election signs (or simply took them down), or refused to shake my hand when I extended it were the exception. The average voter, if there is such a thing, understood that I was simply trying to do my job.

I also learned that an election campaign is demanding. Because it occurred in December and January, I would get up each morning in the dark while my children were still asleep and drive to my first event of the day, often arriving before sunrise. Most nights, I would return home in the dark, well after dinner, when my children were already asleep.

One other thing I learned is that it wasn't easy for me to ask for money. It ran against my nature. However, every politician needs to solicit donations and that means asking people directly. Over the years, I have become more comfortable with this, knowing it's a necessity.

Election night, on January 23, 2006, was tough for me and my family. Pam, my mother, my two brothers and their wives, and my daughter and her husband had all travelled to be with me at my campaign headquarters. Because I was a star candidate (that cursed word again!), the media were also there in large numbers. When the results were announced, I learned that I had been decisively beaten by over nine thousand votes, although not humiliated. Almost eighteen thousand people had voted for me, for which I am most grateful.

"How does it feel to lose?" people kept asking that night. Absolutely awful was one answer. More generously, I tried to see it as constructively as possible—I had learned a lot, which is true. But I am competitive, so it took a little time for me to really see the upside of the experience. Another question I got asked a lot that evening was about my future plans, to which I really didn't have an answer, except to say that I would not rush my decision. I needed time to figure things out. Resuming my job at the CSA was not an option, given that the Liberals were no longer in power and that I had openly demonstrated my allegiance to them. I was now unemployed. Thus began a short period in my life when I would work in the private sector. In some ways, it felt exhilarating. If I did decide to run again, I knew it wouldn't happen for at least two years, given that Prime Minister Harper had won a minority government.

Shortly after my defeat, the National Speakers Bureau asked me whether I was interested in becoming a paid public speaker. I accepted their offer and over the next two years spoke at more events than I can count. Typically, my talks were about my space career. I had of course done this many times before, but the difference now was that I got paid for it. Occasionally, I was also hired as a motivational speaker, something I enjoyed doing.

Around the same time, I was invited to join the board of directors of an oil sands company called UTS Energy, which was actively exploring, although not producing. I was surprised by the invitation, given that I had been speaking publicly for many years about the importance of protecting our planet. All the same, I wanted to better understand this economically important sector, particularly its environmental aspects—and whether there was room for improvement. After much thought, I accepted the invitation.

Those opposed to the development of the oil sands often called them tar sands and made sure the world saw photos of the large tailings ponds, especially when birds had tragically landed in them. It was clear that broad resistance was building, particularly among young people worried about the future of their planet. The arguments, at least on the environmental side, were compelling. There was also no argument with the fact that the oil and gas sector was the largest producer of greenhouse gases in Canada.

I came to realize that it was not a simple choice between closing down everything right away or doing the opposite and allowing unfettered growth. It was more complex than that. A fact all too often dismissed—an inconvenient truth, to quote Al Gore—was that the oil sands were contributing in a major way to Canada's wealth, as millions of barrels were exported to our southern neighbour. There was another reality: even as renewable forms of energy were beginning to gain traction, the need for oil was not going to disappear overnight, whether for fuel or for the many products of the petrochemical industry.

That said, there was an urgent need for improvements in production to reduce greenhouse gases, use less energy, and better manage waste products. The main question for me was whether those improvements could be made quickly enough to allow oil to be developed more sustainably. At the time, oil extraction using the steam-assisted gravity drainage method was an encouraging development but only on a limited scale. The upside of it was that it didn't disfigure the landscape to the same extent as open pit mining. The downside was that producing steam was energy intensive. Carbon capture and sequestration was also being explored, but as yet it was far from proven as a viable and cost-effective solution for carbon storage.

The two years I spent with UTS Energy provided me with an invaluable education, allowing me to understand an issue that has only grown in importance. I have sympathy for this sector and recognize the wealth it has created for our country, but at the same time I also believe it has no choice but to commit itself in a systematic way to the "net-zero by 2050" goal that Canada has set itself. The writing is on the wall. That means a reduction in greenhouse gases. There are no silver bullets that will allow us to continue on our current trajectory.

In the months that followed my election defeat, I had time to reflect on what had happened beyond my own loss. Despite Canada's economy being in good shape, the Liberal Party had suffered a major defeat and it would need to seriously examine itself as it began a rebuilding process that would take time. Canadians had clearly told us that they wanted change. We were a spent force. The questions in my mind were this: Did I want to participate in the undoubtedly long process of rebuilding my party, a party I had joined barely two months before? Did I have the patience to commit to something that would no doubt require many years, or did I want to close the door on politics and move on to something else?

You have probably guessed by now that it's not in my nature to simply walk away after a loss. I didn't get to my previous stations in life because I took the easy (or any other) way out. Also, despite my defeat, I felt a quiet pride at having stepped up to the plate. Unlike those who sit on the sidelines and criticize those in public office, I had acted on my beliefs. I had told Canadians where I stood and accepted their verdict and the consequences for my life.

In the ensuing months, I became increasingly engaged with my party. I made it clear that I wanted to be part of the renewal process, knowing full well there was a lot of work to do and that it would take time. I was in it for the long haul. Despite the state of the Liberal Party then, I believed in its values and its positions on key issues, and I took pride in its long and successful history in building this country. I wanted to be part of its next chapter, and to help see it forward.

One opportunity for renewal was a policy convention of the Quebec wing of the party in Montreal in the fall of 2006, where new ideas were presented and debated. I regard one policy initiative I introduced with two colleagues, Fabrice Rivault and Hervé Rivet, as significant: a resolution recognizing Quebec as a nation. This was a controversial stand, and some senior Liberals were not pleased. Senator Serge Joyal, for whom I have great respect, was furious and told me so, but the three of us went ahead anyway. Being a Quebecer, and having fought on the political front lines of the issue, I believed it was essential for two reasons: first, to recognize something that was important to most Quebecers, an acknowledgement of their distinctness; and second, to take the issue off the table, in effect to neutralize it. Michael Ignatieff, a candidate for the leadership of the party, supported the resolution when I presented it, and it was adopted by more than 80 per cent of members. A month later, the Bloc tried to exploit it for its own purposes but was outflanked by Stephen Harper, who had a modified version adopted by Parliament in late November, specifically, "That this House recognize that the Quebecois form a Nation within a United Canada."

Notwithstanding the political chess-playing by the different parties, I felt I had accomplished my objective. For Quebec nationalists, identity will always be important. It drives everything. The motion adopted by Parliament acknowledged the desire of many Quebecers to be recognized as different, but did so without provoking a stronger desire to go it alone. It lowered the temperature.

In December 2006, the Liberals selected a new leader, Stéphane Dion. The leadership convention in Montreal was an energetic and raucous affair that attracted three thousand Liberals and eight candidates.

Stéphane was not the early favourite but won on the fourth ballot because he and his team were effective in rallying support from other camps. He had focused on what he called the Green Shift—the need to address climate change. He was viewed as a credible champion of the environment, having played the important role of chairing the United Nations COP11 conference in Montreal. The Green Shift included an energy tax on carbon fuels, which economists agreed was the best way to incentivize people to lower their own consumption. Less well known to the public, the Green Shift was also intended to be revenue neutral because the money collected from taxing carbon would be used to reduce people's income taxes and increase family support payments. This plan would be put to the test in the 2008 election.

In political life, it's not unheard of for people in the same party to hold different views. Stéphane Dion, an expert on Canada's Constitution and the father of the Clarity Act (the law setting out the conditions for a province to leave Canada), was not pleased that I had co-authored the resolution recognizing Quebec as a nation, and he told me so in no uncertain terms. I replied that you can't make an omelette without breaking eggs, knowing full well that my future in the party was no longer assured, given that Stéphane was now the leader.

In fact, the outlook appeared rather bleak. In early 2007, I had expressed an interest in running in Outremont when Jean Lapierre retired—a non-starter as it turned out, as Stéphane chose Jocelyn Coulon. Later that year, I made it known that I wanted to run in Westmount–Ville Marie, where I lived, when the sitting MP, Lucienne Robillard, announced that she would retire. Initially, Stéphane did not support this, and I won't speculate as to why. Whatever the reason, I was at a crossroads and was left with no choice but to leave political life, which I then made perfectly clear in an interview with journalist Elizabeth Thompson published in the *Montreal Gazette*. Shortly after it appeared, Stéphane invited me for dinner at Stornoway, the official residence of the leader of the Opposition, where we discussed my future and he asked me to become the candidate for Westmount–Ville

Marie. Simply put, he didn't want me to leave politics, expressing his belief that I could make an important contribution.

Although my relationship with Stéphane was initially bumpy, I have the greatest respect for him—he is a man of powerful intellect, of ethics, and of conviction, who was always motivated by the highest ideals. I consider him a friend to this day.

It goes without saying that anyone who enters politics should be guided by a moral compass. You need to answer the question: Why am I doing this? Unfortunately, if they were to answer that question honestly, more than a few politicians would have to say that it's about acquiring power and then doing whatever it takes to hold on to it. However, I would like to think that most politicians, myself included, go into the profession to do what they believe is right for the country, while also recognizing that what they believe in may indeed be what gets them kicked out.

Politics boils down to making decisions. I'm a Trekky at heart, particularly a fan of Mr. Spock. Why? Because Spock bases his decisions on logic. Yet I also recognize that logic by itself does not always lay the groundwork for the best decisions. Human emotion, seen by Vulcans as a flaw, needs to play into it, and it was often his emotions that allowed Captain Kirk to prevail over his adversaries. So, while others may not agree with my views, they should at least know that, without being rigid about it, I try to be scrupulous in applying logic to my own decision-making.

During the summer of 2007, I took a short break from politics, and our family flew to Switzerland to visit Yves, his wife, Carina, and their newborn son, Elliott, my first grandchild. This was an important milestone in my life. Becoming a grandparent is one of those transitional moments when you realize a generation beyond that of your children has now entered the world. I held Elliott in my arms and felt as close to him as when I had held his father as a baby. At some primitive level, that sense of continuity makes you feel good.

We all tend to live in the immediate. The arrival of a grandchild forces us to look at the continuum of life and at our place in it. It provides perspective.

On October 19, 2007, Stéphane Dion announced that I would be the Liberal candidate in Westmount–Ville Marie. I would run in a by-election the next summer, one of three in the country. While everyone said this was a safe Liberal riding, I was taking no chances. As far as I was concerned, nothing was safe, especially in the shifting political landscape, and I had yet to prove myself as a successful candidate. For those reasons, I wanted to do more than win. I wanted to win convincingly.

I had learned a great deal since my defeat in 2006. I was no longer a rookie with mere weeks to prepare; this time, I was ready. I would be running in the riding where I lived and had no excuse for not knowing the issues of importance to my constituents. I organized my campaign team months in advance. I asked Hervé Rivet, my campaign manager from 2006, to lead my team once again, and he accepted, wanting to avenge our loss in 2006.

The by-election kicked off on July 25 and was set to last forty-five days. My team hit the ground running. Since there were only three by-elections underway in the country, I received many offers of assistance from sitting Liberal MPs, which I accepted gratefully. Even Ken Dryden offered to knock on some doors with me in Notre-Dame-de-Grâce, and I was delighted to receive his assistance. As everyone knows, Ken is close to a deity in Montreal, having won six Stanley Cups as goalie of "Les Glorieux." The plan was for me to knock on doors with Ken beside me. As an elected politician, he would be there to help answer some of the questions I might be asked. Our plan was to spend no more than two or three minutes at each door. It was a great plan, in theory.

That's not exactly how it unfolded. At the first home, the owner opened the door, looked at me and then at Ken, and a wide grin appeared on his face as he shouted back into the house: "Honey, Ken Dryden is at the door, get the kids and bring the camera!" Then came the pictures. This was followed by a trip down memory lane to the glory days of Les Canadiens. Ken would dutifully answer every question that came his way as I faded into the background. Periodically Ken would try to introduce me as the candidate, but to no avail. I felt like the kid at the back of the class frantically waving his hand to get the teacher's attention. But this being Montreal, with Ken Dryden standing on the front steps, my constituents only wanted to talk hockey, and Ken, being the polite person that he is, was not good at ending a conversation. The result was that we only made it to six doors that evening, although the news that I had brought Ken Dryden to the neighbourhood spread like wildfire, to my ultimate benefit.

As my campaign progressed, rumours began to circulate that Prime Minister Harper, hoping for a majority, might call a general election in September, and this is precisely what happened. In fact, the election was called one day before my scheduled by-election. It would take place on October 14, extending my campaign to eighty-two days. Because most of my original team of volunteers had committed to helping me only until September 8, I had to scramble to find new volunteers to take me across the finish line.

My main opponent was the NDP candidate Anne Lagacé Dowson, a well-known radio personality. Given her experience as a talk show host, she proved to be a skilled communicator as we sparred in the many public debates organized during the campaign. One day, when I was campaigning on Sherbrooke Street, I noticed that Anne, accompanied by Thomas Mulcair, then NDP MP for Outremont, was doing the same. Unbeknownst to me, Pam was walking down the street to do her shopping when she was intercepted by Mulcair, who asked her if he could introduce her to the NDP candidate and then began to expound on my apparent shortcomings. Pam, who usually displays great self-control, informed him who he was talking to and then tore into him. Looking sheepish and turning red, Mulcair told her that it was all just part of the political game and then inquired solicitously about me. Pam was not amused, although I got a kick out of it.

On election day, I won with 46.5 per cent of the vote, versus 23 per cent for the NDP, a sweet and satisfying moment for me. Notwithstanding my own victory, the Liberal seat count was reduced from 103 to 77. Clearly, we had not presented a sufficiently compelling case to be returned to government. Did our leader fail to connect? Was the memory of the previous Liberal government still too fresh in people's minds? Did we lose because Canadians didn't understand the Green Shift or weren't ready for it? Did Canadians feel the current government was better prepared to handle the global economic downturn? Whatever the reason, Stéphane Dion took it hard, stepping down as leader. Michael Ignatieff would succeed him, first on an interim basis and later as the official leader.

As for myself, I was glad that two and a half years of hard work and perseverance had paid off. I had redeemed myself and would now get to sit in the House of Commons with other Liberal MPs and serve as a member of the official Opposition. I looked forward to learning my craft as a legislator and as the critic for industry, science, and technology. I took the long-term view that starting in opposition was not a bad way to learn my job and gain experience. The Liberal Party, though, faced some serious challenges.

TEN

EVEN THOUGH I SAT in the last row of the opposition benches, I was now one of 308 MPs of Canada's fortieth Parliament, responsible for passing the federal laws of our country—laws that could affect the life of every Canadian. The importance of this was driven home for me the first time I rose to have my vote counted. It was sobering to realize I would have to publicly declare my beliefs for all to see and spar with people who didn't share my point of view. My job was to persuade them, hopefully in a civilized manner, that my views were the right ones.

As with most things, in time I began to feel more comfortable. I was gaining experience, taking part in debates, and questioning ministers in the daily back-and-forth known as Question Period. My goal was to ask my question (and later, as a minister, to answer questions) without reading from crib notes. This is harder than you might think and must be accomplished in less than thirty-five seconds, before your microphone is cut off. It's also much harder if you are being heckled while you're speaking. It's easy to understand why some people stumble. Many an opposition MP has begun asking a brilliant question with a riveting preamble, only to be cut off before getting to their actual question. Timing yourself is critical. While asking a probing question is a skill that all MPs can learn in due course, what is shocking to learn, as a new MP, is that the government minister responding to your question is under no obligation to answer it, or at least to answer it directly. Fortunately, the media are there to point out when the government is being purposely evasive.

MPs are expected to give speeches supporting or criticizing proposed legislation, and when you're in opposition you don't have the luxury of having a prepared speech handed to you. You write your own, and in my case, it had to be bilingual. I would sometimes get up at three in the morning to write them. You were expected to speak for twenty minutes or, if you shared your time with a colleague, ten minutes. That can be a long time to deliver words of substance. There is also a skill in pacing yourself so that you don't have to rush frantically to finish when the Speaker gives you the "one minute left" signal. An experienced speaker will finish with a memorable concluding remark and will rarely consult their notes as they speak, preferring to make eye contact with the Speaker and other MPs. I remember admiring the way Bob Rae would get up and speak eloquently and extemporaneously for twenty minutes with only a small scrap of paper containing a few scribbled notes to guide him. For me, he was the gold standard.

In addition to my duties in the House, I also sat on the parliamentary Standing Committee on Industry, Science and Technology, which, as its name implies, focused on legislation or special studies related to industry, science, and technology. For me, the key to being a good committee member was to know the subject matter. Nothing is more effective in politics than knowing what you're talking about.

As I continued my apprenticeship throughout 2009, two wonderful events would occur in my life: the arrival of Yves' second son, Emil, followed closely by Simone's daughter, Ela. And just like that, I was now the proud grandfather of three children.

As interesting and exciting as it may appear, the life of an MP can also be lonely, for you are separated from your loved ones for long periods of time. I was luckier than most because my widowed mother lived in Ottawa. One day, she suggested I move back in with her. She lived a brisk thirty-five-minute walk from the Hill. This meant my own bed and home-cooked meals, something I could not pass up because my mother was an exceptional cook. I enjoyed living with her for the next four years whenever I was in Ottawa. She loved politics, and every evening we would discuss the day's events. I cherish those times, which allowed me to get to know her much better, especially since I had left home at such a young age.

During our time together, I learned so much about her, including details of her life that she had never shared: how she grew up in the Depression and how the family had to keep moving as her father looked for employment, how she and her five sisters had all trained as nurses at the Jeffery Hale Hospital in Quebec City, and the circumstances in which she met my father. Getting to know her better, I realized just how smart she was, something I had not recognized as a child.

It was during this time that I urged her to put some of her early life on paper, particularly the period when she was growing up. She was reluctant to do so, but I did manage to coax some handwritten pages out of her, which I cherish today. I took my own advice and have done the same for my own children, making a point now and then to jot down some of the early details of my life, something that I later resolved to flesh out into some kind of memoir.

Of course, nothing lasts forever. In 2013, feeling less capable of living on her own when I wasn't there, my mother moved to Montreal to be closer to more family, including mine.

Occasionally, MPs who are not ministers can make an important contribution to Parliament by presenting either a motion or legislation known as a private member's bill. In both cases, Parliament votes on what is presented. These opportunities are rare in the life of a backbench MP and depend on a random draw at the beginning of each new Parliament. If yours is one of the last names to be drawn, it's doubtful you'll get to present your motion or bill, so there is an element of luck.

I was fortunate that my chance came up on June 11, 2009, when I introduced a bill to create a permanent national commissioner for children and youth. The idea had been suggested to me by a constituent, Ginette Sauvé-Frankel, and I found it compelling. Canada had ratified the United Nations Convention on the Rights of the Child in 1991 but had not followed up with a mechanism to monitor whether it was living up to its obligations. More than sixty other countries had created a children's commissioner, but not Canada. Senator Landon Pearson (the daughter-in-law of Prime Minister Lester B. Pearson) had worked tirelessly for almost two decades to promote the idea, and I wanted to continue her work. I subscribed to the belief, expressed in a UNICEF report ("Child Poverty in Perspective: An Overview of Child Well-Being in Rich Countries," Innocenti Report Card 7, 2007, UNICEF Innocenti Research Centre, Florence), that "the true measure of a nation's standing is how well it attends to its children—their health and safety, their material security, their education and socialization, and their sense of being loved, valued, and included in the families and societies into which they are born."

What motivated me the most in presenting my bill was the realization that the seven million children of Canada were largely voiceless. Adults spoke on their behalf on all matters, the assumption being that adults knew best. I was supported in my efforts by UNICEF, the Canadian Paediatric Society, and many well-established children-focused non-governmental organizations, as well as several provincial child advocates who recognized that the federal government had important responsibilities with respect to children. Certain federal laws determined how children were treated—for example, criminal law dealing with juvenile delinquents as well as laws pertaining to marriage and divorce. And, of course, the federal government had direct responsibilities for Indigenous children. Sadly, while my own party supported it, my bill was defeated by the Conservative majority. They argued that the government did not need yet another commissioner checking up on them and, more importantly, that parents didn't need to be told how to raise their children.

This setback did not diminish my resolve. I would present my bill again during the forty-first Parliament, but, once again, it was defeated. At that point, I decided to put it on hold and hope for the necessary support from a future Liberal government.

In March 2011, sensing he could achieve a majority, Prime Minister Harper set the stage for an election. He did this when he tabled a budget none of the other parties found acceptable. As he was leading a minority government, he needed some opposition support when the time came to vote. He argued that he had included many concessions to bring the NDP onside (there is some truth to this), but in the end, Jack Layton said no. Since a majority vote against a budget indicates non-confidence in the government, it triggers an election. The governor general dissolved Parliament and election day was set for May 2. It would prove to be a case of "third time lucky" for the Conservatives. More significantly for my own party, it would be the first time in the history of Canada that we were relegated to third-party status. We would hit rock bottom, with only thirty-four elected members. I was one of those, hanging on by the skin of my teeth.

There were several reasons to explain this outcome, but two stand out. The first was Jack Layton's "orange wave," and the second was party leader Michael Ignatieff's inability to overcome the lethal Conservative attack ads pointing out that he had lived outside the country for thirty-five years and was therefore "not there for Canadians" or "just visiting."

As the campaign moved into high gear, it became clear that Layton was gaining the sympathy of Quebecers for several reasons. He was from Quebec, having grown up in Hudson, a predominantly English community just west of Montreal, and he spoke French, which Quebecers appreciate, even if you make mistakes (they admire you for trying). Also important, he was personable and had a way of connecting with whomever he was talking to. Finally, because he was fighting cancer, he earned the admiration of many for his courage, stamina, and passion throughout the campaign. For all those reasons, the NDP became an unstoppable force in Quebec, gathering momentum as the campaign progressed. As each day went by, projected Liberal losses increased and even my own riding of Westmount–Ville Marie was predicted to fall to the NDP.

Although I didn't personally know him well, except in his public role as leader of the NDP, let me say this about Jack Layton. What he accomplished in the 2011 election, particularly in Quebec, was remarkable, and I regret that he did not have the opportunity to lead his party in the next Parliament, a job he deserved. He brought a civility to the often raucous exchanges in the House of Commons. It would have been fitting to see him take his place as leader of the official Opposition. Sadly, I would join other colleagues at his funeral later that summer, at a moving ceremony in Toronto.

Momentum affects morale. My own team was certainly feeling the momentum of Jack's orange wave as we soldiered on bravely, trying to reassure ourselves that we could still pull it off on election night. Constituents were telling me they were still sure I would win but that this time they might vote NDP to show that party some support. That's how much sympathy Jack Layton had attracted. Naturally, I was concerned that if too many of them switched their vote "just this once," I would be defeated, but there was not much I could say. As it turned out, my concern was not misplaced.

On election night, as was becoming the custom, I sat at home with my family waiting for the results. As the returns started coming in, it was clear that the NDP was about to score a major upset. Ultimately it won fifty-nine of Quebec's seventy-five seats and as a result became the official Opposition, with the Conservatives leading. Liberals, on the other hand, would manage to hold only seven seats in Quebec, down from fourteen. As the numbers continued coming in and I continued to trail, Radio-Canada declared an NDP victory in my riding. The computer algorithms had spoken, and their predictions were invariably right—or at least nineteen times out of twenty. I was disappointed, but not surprised.

Knowing the media were waiting to hear from me, I contacted François Rivet, my campaign manager (Hervé's brother), and told him I was ready to concede. He advised me to wait, saying that the advance polling votes had not yet been counted and that things could change. (Advance polls are counted last.) I told him I thought it would look bad if I delayed any further, given that it was already close to 11 p.m. Pam and I walked over to the campaign office, and before a tearful group of volunteers, I conceded defeat to the NDP candidate Joanne Corbeil and wished her good luck. I then thanked my team and returned home, completely drained. Pam and I went to bed emotionally and physically exhausted, and fell asleep quickly.

At two o'clock, the phone rang and I was informed that the final numbers had been tallied and that I had won by 642 votes. Take that, Radio-Canada! That's the last time I would put my trust in an algorithm rather than my campaign manager. It turned out that advance voters in my riding were mostly diehard Liberals, come hell or high water. In addition, their votes were cast before the orange wave had truly gathered momentum. I leapt out of bed and woke the rest of the family to tell them the good news. I felt like dancing in the street and may have done so in my bathrobe. I would not be able to sleep the rest of the night and was back at my campaign office the following morning to help with the cleanup. As it happened, the volunteers who were already there had not heard the final result and they greeted me with long, tired faces, probably wondering what the big smile on my face was all about. Despite the devastating results for our party, it was such a pleasure to share the news of my belated victory with them.

I had now campaigned in three elections, losing the first, winning the second, and surviving the third. Politics was certainly a rough business, where you couldn't take anything for granted. Would I ever manage to be on the winning side? Being an eternal optimist, I definitely felt it would happen one day. As I helped clear out my campaign office, I recommitted myself to that goal.

First, though, my party needed to dig itself out of an exceedingly deep hole, a daunting task requiring bold change as we prepared to take on our new, never-before-assumed role as the third party.

ELEVEN

LET'S BE BLUNT: with only thirty-four seats and 19 per cent of the popular vote, the 2011 Liberal Party results were disastrous. Never in the history of the country had we come third in an election. If we thought 2008 had been bad, this was nothing short of calamitous. Some pundits were even predicting the end of the party. Five years of effort to rebuild had been for naught. If anyone had clung to the notion that Canadians would see the light and re-elect a Liberal government after a few years in purgatory, the 2011 election was a massive cold shower. It was time for a heavy dose of humility and a radically new approach.

As we returned to Parliament, I could not help but notice that the entire Liberal caucus—which at the time still included senators—now fit in a much smaller conference room and that the seven remaining MPs from Quebec could meet in anyone's office. We had lost many good people and now, as the third party, we would have far fewer resources at our disposal. This forced us to take a more informal approach as we contemplated, once again, the process of rebuilding. A leadership race would also have to be called.

After the dust settled, the caucus held a vote to pick an interim leader, pending the outcome of the formal leadership race. Although more than one name was mentioned, only Bob Rae seemed willing to step up to the plate. Fortunately, he had the experience and skills for the job. And yet, in terms of optics, it struck me that it would look better if the public were to see a race rather than a coronation. So I threw my hat in the ring, having no illusions about the outcome. Before the vote, the party had stipulated that whoever won, that person could not later run to become party leader. Bob and I were comfortable with this. The party's constitution was also amended to defer the start of the leadership race to November 2012, with the selection of the new leader in April 2013. It was important for us to take the time to get organized, and besides, there was no rush, given the Conservative majority.

As expected, Bob Rae was chosen as interim leader, and he made me his House leader, a job that required me to become more knowledgeable about House procedures and standing orders.

Next on our agenda were the preparations for a Liberal convention, to be held in Ottawa in January 2012. We needed to meet as a party and do some serious soul-searching, although after our crushing defeat, some wondered whether anyone would even show up. It would be a test of whether the Liberal heart was still beating. Lo and behold, thousands came out, and that fact alone lifted everyone's spirits. Despite coming third, Canadians still believed that there was a future for our party—provided, of course, that major changes were implemented.

In an effort to widen the tent and renew ourselves, one of those changes was the creation of the Liberal "supporter" category, allowing anyone who became a "supporter" to vote in the future leadership race without needing to become a paid-up party member.

It was also at the Ottawa convention that the party voted in favour of legalizing marijuana, a daring policy initiative that would generate a great deal of discussion across party lines—with some for, some against, and some in favour only of decriminalization—but that in the end would be implemented without much fuss.

This was the convention where Bob Rae, Scott Brison, John McCallum, Dominic LeBlanc, Carolyn Bennett, and I performed a skit for *This Hour Has 22 Minutes* showing us holding our national convention in a restaurant lunch booth because we had fallen on hard times. Mark Critch interviewed us, and we had great fun. Bob Rae couldn't stop giggling, Scott Brison accused me of putting my fries on his tab, and I drank Tang and talked about space. When you're down and out, it helps to have a sense of humour.

Throughout the spring of 2012, I pondered whether I would run for the party leadership. Justin To, a friend with whom I had worked when he was in the opposition leader's office, had floated the idea with me. It was by no means an easy decision; in fact, I wondered whether I was crazy to even entertain the notion. Was this the path I really wanted to take, with the accompanying level of scrutiny, criticism, and exposure?

On a more practical level, you needed a campaign team and that team needed to be more than just enthusiastic. It needed solid experience running a national-level campaign. It also needed money. You had to raise $75,000 just to enter the race and you would need a lot more to conduct a full campaign. Also, if you spent more than you received in donations, you had to pay off your debt within a specified time or risk penalties from Elections Canada.

Unlike a riding election, which only involves your constituents and lasts five to seven weeks, a party leadership campaign spans the entire country and lasts five months. Still, with that in mind, I decided to put out feelers to see if there was any interest in my running in the first place. The question I needed to answer was whether I could assemble a credible team and enough supporters willing to fund me and get behind the idea of me as party leader.

I also needed to answer a more fundamental question: Did I have the necessary qualifications to become the leader and potentially, one day, the prime minister? Before taking such a serious leap, I needed to believe that I was a suitable candidate for the job. On the positive side, I had a strong work ethic and considerable experience leading people. I also felt I had solid policy ideas, a firm grasp of the issues at hand, and a proven track record of accomplishments. Less certain was whether I had the kind of personality that would appeal to voters.

My parents' wedding in Ottawa, 1946. (Photograph courtesy of the author)

In Saint-Jean-sur-Richelieu, 1959: Dad teaching Braun, Charles, and me how to play tennis. (Photograph courtesy of the author)

My brothers Braun, Charles, Philippe, and me in London, England, 1963. (Photograph courtesy of the author)

The *Pickle*, 1969. (Photograph courtesy of the author)

Marching off the parade square at RMC after my graduation, 1970. (Photograph courtesy of the author)

The first Canadian astronauts in 1984, shortly after our selection. (Photograph courtesy of Brian Smale)

The 1983 "Astronaut Wanted" ad that started it all. (Image credit: National Research Council of Canada)

My first meeting with Commander Bob Crippen at JSC in August 1984. (Image credit: NASA)

My grandmother and parents with Jacqueline, Simone, and Yves in Florida, the night before the launch of STS-41G. (Photograph courtesy of the author)

Being presented in the House of Commons with my family after STS-41G, 1984. (Photograph courtesy of the *Canadian Press* /Andy Clark)

At KSC, training with the STS-77 crew: pictured on the launch pad with the Shuttle behind us. (Image credit: NASA)

Operating the Commercial Float Zone Furnace experiment on STS-77. (Image credit: NASA)

Meeting Ronald Reagan in the Oval Office with Commander Bob Crippen looking on, 1984. (Photograph courtesy of the White House, Washington, D.C.)

At the Rockcliffe Flying Club with Bjarni Tryggvason, who taught me how to fly, 1985. (Photograph courtesy of the author)

Shaking hands with Wayne Gretzky before the 1985 NHL All-Star Game at the Calgary Saddledome. (Photograph courtesy of the author)

All children love space and the planets. (Photograph courtesy of the author)

Preparing to capcom STS-67 in JSC's original Mission Control Room, 1995. (Image credit: NASA)

In the early morning of May 19, 1996, as STS-77 (*Endeavour*) begins its journey to space. (Image credit: NASA)

My four children with Pam in Florida the night before my launch on STS-97. (Photograph courtesy of the author)

Family portrait taken in late 1999, with Pamela, Adrien, and George, as I began training for STS-97. (Image credit: NASA)

In the white room on the launch pad, prior to entering *Endeavour* for STS-97: the side hatch door opening is visible behind me. (Image credit: NASA)

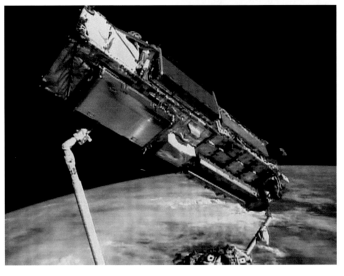

The P6 truss segment after I had removed it from *Endeavour*'s payload bay with the Canadarm, prior to it being attached to the ISS on mission STS-97. (Image credit: NASA)

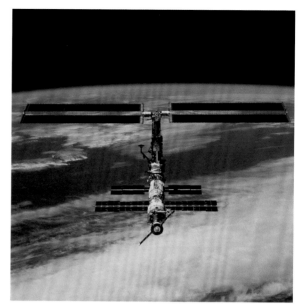

The ISS in December 2000, after we added the first large solar panels on STS-97. (Image credit: NASA)

Pamela and me (alongside Margaret Atwood) meeting the Queen at Rideau Hall during her Golden Jubilee visit to Canada, 2002. (Photograph courtesy of Government House, Rideau Hall)

My family in 2011: my mother, Pamela, Adrien and George, Yves and Simone with spouses Carina and Ozgur, and grandchildren Elliott, Emil, and Ela. (Photograph courtesy of the author)

Debating Justin Trudeau in Mississauga during the Liberal Leadership race, 2013. (Photograph courtesy of the author)

The meeting in October 2015 when the prime minister (in the company of Gerald Butts and Katie Telford) asked me to become transport minister. (Photograph by Adam Scotti, courtesy of the Prime Minister's Office)

Post-retirement, 2023: finally getting around to assembling my own Space Station. (Photograph courtesy of the author)

Pretty much everyone loves and admires astronauts; not so much politicians. The fact is that I was not charismatic. It has never been in my makeup, and to my mind it is something you either have or you don't. So the question was: Would Liberals support a capable and serious person without the sizzle?

Meanwhile, other potential candidates were considering their chances, although it was clear that one of them, Justin Trudeau, a son of former prime minister Pierre Elliott Trudeau, had already made up his mind. His very public charitable boxing match with Conservative senator Patrick Brazeau signalled to Canadians that he was moving into high gear. He had made the decision to run and had lots of people willing to support him. He was young, attractive, and had a last name that resonated powerfully with many Canadians, and all Liberals.

I set my own soul-searching aside for a while and flew to Turkey for a holiday with Pam and the boys, joining my daughter, Simone, and her family. While there, I inadvertently tested the Turkish health care system. One morning, while putting on my shirt, I felt a painful sting on my right side, followed a few seconds later by another behind my right leg. I looked down and saw a scorpion scuttling away. It had been hiding in my shirt. I immediately stepped on it (I was wearing shoes). At this point, I had no idea what might happen. I learned only later that some scorpions are more venomous than others. To be on the safe side, my Turkish son-in-law, Ozgur, drove me to the local hospital, forty minutes away. A scorpion sting is not unusual in Turkey, so a doctor administered a steroid intravenously and I was put under observation for a couple of hours. Apart from some redness at the sting sites, I felt fine and was discharged. My total bill was sixty-four dollars, which is worth every cent after you've been stung by a scorpion, and as far as I was concerned, the Turkish health care system had passed with flying colours. There were of course the usual family jokes, saying that I'd been stung by a radioactive scorpion and that I was now Scorpionman.

It turns out that Justin To had been quite busy while I was away, assembling a team of experienced people who, after meeting me one-on-one, were willing to give me their support. Andy Mitchell, a veteran Liberal minister in both the Chrétien and Martin governments, agreed to be my campaign manager. Derek Ferguson and Anne Dawson, both experts on the communications side, would help develop my campaign strategy. Justin and Peter Harrison would focus on the policy initiatives I would present during the race, and Daniel Langer would take care of all the logistics, such as travel and accommodation, for my cross-country tour. Ali Salam, Jarett Lalonde, and Alain Berinstain would plan my events in Ontario, the western provinces, and Quebec. Jordan Owens would be my press secretary and handle media relations, and Christine Michaud would manage my information database. Jean Proulx would coordinate my schedule, and Zenon Domanchuk would be my financial agent. In addition to having a team, donations began to appear, and it became clear that I would succeed in raising the $75,000 needed to enter the race.

All that was left for me to decide was whether I was in or not. After a lot of introspection, and with Pam's blessing, I decided to take the leap. Was it an act of hubris on my part? Others can opine on that. What I will say is that in making decisions, I tend to weigh the pros and cons, and to then rarely second-guess myself, believing it too often leads to regret. As I saw it, even if I didn't win, running would be a worthwhile experience. It would raise my profile, not only within the party but across the country, and help me further hone my campaigning skills.

I kicked off my campaign in Montreal on November 28. Ultimately, nine of us would face off in five debates across the country, beginning in Vancouver. Each debate attracted national attention and gave Canadians a chance to see the candidates: Justin Trudeau, Joyce Murray, Karen McCrimmon, Martin Cauchon, Martha Hall Findlay, George Takach, David Bertschi, Deborah Coyne, and me. Between debates, we campaigned on our own. The new leader would be chosen on April 14, 2013.

In a campaign, you must regularly offer fresh ideas, which I did. Meanwhile, everyone is asking the same question: What makes you more appealing than the others? It's a question I was also asking myself. As the campaign progresses and the competition intensifies, you're tempted to spend more and more money on publicity and travel. Polls and opinion pieces begin to appear, giving you an idea of how well or how badly you're doing. There is never any downtime, and eventually you begin to sense how it will end.

In leadership races, all the candidates belong to the same party, and that makes criticizing your opponents a delicate affair. Ideally, when it's over, you will all shake hands and continue working together towards a common goal. So there are lines you're not going to cross, even though you are tempted to do so. On the other hand, your mettle is being tested, and it's important for everyone to know whether the future leader has what it takes to win. It's all about balance, but a leadership race can't just be a love-in where you're all saying how great everyone is.

This was at least part of the reason I challenged Justin Trudeau in the third debate. I asked him directly what made him think he was qualified for the job. Of course, I was making the point that an important consideration in choosing a leader is the experience and track record they bring to the table. I also realized, in asking the question, that it was easy to interpret it as not so much a question as a statement: You're not qualified for the job. No matter how it was interpreted, it would no doubt affect the way he saw me from that moment onwards. Nevertheless, it was a fair question, and one that he would be asked many times in the future.

One of the things I admired most about my campaign team was its brutal honesty. Such honesty is not a given in such situations, where delivering bad news can sometimes be met with denial. In early March, one month before the end of the race, they told me I could not catch Justin Trudeau. Our internal polling made that clear. I accepted their verdict, knowing it to be true, and given the need to avoid spending money that would only increase my debt, I announced that I was dropping out of the race. One journalist asked me why I was so sure I could not win. I replied: "I believe in mathematics, and the math tells me I can't win." I had phoned Justin the night before to let him know I would be supporting him. A month later, he became our new leader.

It had been an exciting ride. It helped me grow as a politician and in the process gain a better insight into how people viewed me. It's important to know your strengths and your limitations, and running for the leadership revealed both. Throughout the campaign, I was supported by an exceptional team every step of the way. We had some memorable moments together, but I, and only I, am responsible for the outcome. I was not the favourite choice of Canadians to become the next Liberal leader.

In the months that followed, I focused on repaying my debt. My campaign had spent over half a million dollars, and since I had not received that amount in donations, I had to fundraise. I did more than a dozen events and was gratified by the generosity of so many people, some of whom, I know, had not supported my candidacy. I repaid my entire debt by the required deadline. I was proud of that.

Now that the leadership race was over, I settled back into my life as an MP and awaited direction from our newly elected leader. Everything was in place for our party to begin the task of trying to regain the confidence of Canadians and to demonstrate to them that we had our act together. Time would tell.

In 2013, out of the blue, NASA decided, after retiring the space shuttle, to offer one of the Canadarms to Canada for permanent display at the Canadian Space Agency. I had been informed of this and recommended instead that it be displayed in Ottawa, at the Canadian Aviation and Space Museum, a more suitable place for the public to view this icon of Canadian technology. I had made the suggestion to Christian Paradis, the Conservative minister of industry, science, and technology, and he agreed. Shortly thereafter, the museum announced a date for the official unveiling, but I was not invited despite having operated the arm on two of my missions. I felt this was petty, notwithstanding Conservative ministers making the dubious claim that they had no say in issuing the invitations. This was an example of the side of politics I didn't like.

When Justin Trudeau took over as leader, he named me his foreign affairs critic. Having lived abroad for seventeen years of my life, I was delighted to assume this responsibility. He also asked Gen. Andrew Leslie and me to co-chair a group of experts known as the International Affairs Council of Advisors, which provided advice to the leader on defence and foreign issues such as the rise of ISIS, tensions in the Middle East, Canada's defence procurement, and our response to global security threats. It included distinguished academics, former ambassadors, former ministers, and a small group of Liberal MPs. I learned a great deal from my interactions with them. Justin Trudeau met with us two or three times, asking questions and expressing his own views on a number of the issues being discussed.

In my role as foreign affairs critic, 2014 would be a memorable year because of two trips I made to the Middle East. The first took me to Israel, the West Bank, Jordan, and Egypt in the company of Thaer Mukbel, a friend of mine and a previous advisor to Bob Rae. The second took me to Iraq with John Baird, Canada's minister of foreign affairs, and Paul Dewar, the NDP's foreign affairs critic.

The purpose of my first trip was to better understand the dynamics in the seemingly intractable Israeli-Palestinian conflict. There have been fleeting moments of hope over the decades that a negotiated solution might bring a lasting peace, but that solution has remained elusive, despite the misery caused on both sides.

While I had no illusions about the modest weight that Canada carried in the region, I believed it was important for us to play a constructive role. Our friendship with Israel goes back to its creation. We have always believed in its right to exist and that includes defending its territory as any sovereign state would. We also believe in a two-state solution, and that belief is unwavering. Palestinians have the right to exist as a state in the region, next to Israel. That said, several obstacles have thus far stood in the way, chief among them the fact that there are players on each side that are adamantly opposed to a two-state solution.

The Palestinian faction Hamas, a terrorist organization controlling Gaza and strongly supported by Iran, does not recognize Israel's right to exist and is committed to its destruction. It only recognizes the use of violence to achieve its objectives. It has attacked Israel on multiple occasions (in a most brutal and barbaric fashion on October 7, 2023) and will continue to do so for as long as it maintains its control over Gaza. Meanwhile, Israel continues to annex land in the West Bank for settlements.

The task of finding a solution is daunting. Both sides have hardliners who are unwilling to compromise. In addition to agreeing on boundary lines and resolving the issue of settlements, other obstacles that need to be resolved include the fate of Palestinians wanting to return from refugee camps and the fact that Palestinians want East Jerusalem to be their capital. Attempts to broker an acceptable solution have not succeeded because of entrenched positions. And yet, because there are sometimes glimmers of hope that a solution can be found, it's important for a country like Canada to do what it can to encourage dialogue between both sides.

It's also important to recognize that what happens in this region has a spillover effect in Canada, as evidenced by the alarming growth in anti-Semitism and Islamophobia we are witnessing on our own soil, including in my own city of Montreal. Both forms of intolerance may go back centuries, but the continued tensions in the region exacerbate the situation, particularly when Hamas attacks and Israel retaliates.

As critic for the third party, there were no funds to pay for my trip and there was no guarantee that the people I wanted to meet would agree to meet me. In the end, I decided to finance most of it myself and hope for the best. My expectations were exceeded.

I landed in Amman, Jordan, and drove to Jerusalem the next morning, crossing the historic Allenby/King Hussein bridge that spans the Jordan River, almost a creek by Canadian standards. Taking this road trip was an education in itself, helping me to better understand the geography of the region—a region of historical significance that I had learned so much about as a child, as part of my religious education.

This would be my third visit to Israel. I had previously visited as president of the Canadian Space Agency, seeking to create linkages with Israel's impressive space program. I had even spoken to a class of young students working on a microsatellite project.

My purpose on this visit was to hear from Israeli officials about the current tensions in the region and whether hope for a two-state solution was still alive. Despite my third-party status, I was well received. Officials briefed me on Canada–Israel relations, Iran's nuclear program, and various regional issues, including the war in Syria. I also met with Yuval Steinitz, the minister of intelligence, strategic affairs, and international relations. On the possibility of a two-state solution, the people I spoke with were non-committal, while also not rejecting it.

Because I had known Ilan Ramon, the Israeli astronaut who was tragically killed in the *Columbia* shuttle disaster, I took time to meet with his wife, Rona, and visit a site where trees had been planted in his memory. (Ilan had flown over Israel during his mission and expressed the wish to see more trees planted.)

Finally, I took some time to visit Yad Vashem, the World Holocaust Remembrance Center, a deeply moving experience. I viewed one exhibit after another, tracing the events that culminated in the extermination of six million Jews and other earmarked groups, and left deeply shaken and profoundly saddened by what I had seen. When I was a young boy, my father had explained the Holocaust to me, after I had accidentally come across some photos showing dead bodies piled on top of one another in one of the concentration camps. It was the first time in my life that I had seen something that my young mind could not comprehend. How could we inflict such horrors on our fellow humans with the single-minded purpose of exterminating an entire people? I would return to Yad Vashem on subsequent trips to Jerusalem.

To fully appreciate the geopolitics of the region, it is essential to understand conditions in the West Bank, my next destination. This would be my second trip to Ramallah, where I had been invited to appear on a morning radio talk show, followed by a visit to two Palestinian schools, where I spoke to the children about space. The first school was a private one and the second was a public elementary school where I spoke outside, under an awning with a makeshift screen on which to project slides. All children are wide-eyed when hearing about space, and these children were no different. They asked the same questions children everywhere ask. It drove home to me the truth that every child starts out the same way, curious and totally innocent.

Mahmoud Abbas, head of the Palestinian Authority, had agreed to meet me during my visit but was called to London the night before at the request of U.S. secretary of state John Kerry. (I would meet Abbas many years later after I became foreign affairs minister.) Nevertheless, I met with some of his officials, although it was more of a courtesy call than anything else.

I was also able to visit Bethlehem and then Rawabi, the ultra-modern planned city under construction north of Ramallah and financed primarily by Qatar. To cap it off, I joined a group of young Palestinian entrepreneurs for dinner and a frank discussion of the opportunities and obstacles they faced as they tried to develop their businesses. I found this encounter to be particularly valuable. It allowed me to better understand the hopes and aspirations of at least some of the new generation of Palestinians.

Given Egypt's influence in the region, I next flew to Cairo, where I met with officials, including Sameh Shoukry, who would soon be named Egypt's foreign affairs minister. I took the opportunity to raise with them the issue of two prisoners: Canadian journalist Mohamed Fahmy and human rights activist Khaled Al-Qazzaz. Despite Canada's efforts on his behalf, Fahmi was imprisoned in Cairo's notorious maximum-security Scorpion Prison, accused of conspiring with a terrorist group while reporting for Al-Jazeera. He would eventually receive a presidential pardon and be released in September 2015, after spending 438 days in captivity.

I also met with Cairo's Coptic Christian community, a religious minority that has faced discrimination and persecution in Egypt over the centuries, leading some to move to other countries, including Canada. I had previously met with members of a Coptic community in Etobicoke and wanted to better understand the situation of Copts in Egypt. My assessment after meeting with some of their leaders in Cairo was that simmering tensions very much remained.

During my stay, I had the pleasure of meeting staff at the Canadian Embassy and having dinner with the ambassador. I appreciated the courtesies extended to me, especially since I was not even the critic for Canada's official Opposition but rather the critic from the third party. The visit gave me insight into diplomatic life in a country experiencing unrest, allowing me to appreciate the stressful conditions under which our diplomats were living, Cairo having witnessed repeated violence and unrest in the period leading up to General el-Sisi's takeover earlier that year.

Before leaving, knowing this might be a once-in-a-lifetime opportunity, I made a quick visit to the Great Pyramid of Giza, where I even managed to sit on a camel without falling off!

I then returned to Amman, my starting point, where I wanted to speak to officials and gain a better appreciation of the historical role Jordan had played in the region. I would meet the prime minister, who thanked Canada for our steadfast friendship and support. Most memorable for me was my trip to Za'atari refugee camp, which had been running for many years and was well organized. It held about a hundred thousand refugees, mainly Syrians who had fled Bashar al-Assad's brutal regime and the equally brutal advances made by ISIS. I was given a tour of the camp that allowed me to see a food distribution centre operated by the World Food Programme, a classroom where schooling was offered to children, and a medical clinic able to deal with basic needs. One dirt road in the camp was bordered by more than a thousand small shops run by enterprising Syrians and was nicknamed "the Sham-Élysées."

Za'atari was the first refugee camp I had ever visited, but it would not be the last. Meeting refugees and witnessing their living conditions was eye-opening. These were people whose lives had been upended and put on hold indefinitely, people who survived on the dream of a better life for themselves and their children. What was undeniably clear was that Jordan, like Lebanon, was carrying a heavy burden in the number of refugees it was hosting.

Overall, this ten-day trip gave me a greater appreciation of the complex dynamics of the region. Sadly, deadly conflict would erupt two months later between Hamas and Israel, reigniting tensions in the region and leading to thousands of casualties.

In early September, I embarked on a second Middle East trip, this time to Iraq. John Baird was going to Baghdad and Erbil and offered to take his two critics with him: Paul Dewar and me. Both of us welcomed this opportunity.

ISIS was occupying parts of Iraq, including the city of Mosul. They were also setting their sights on Erbil in the autonomous Kurdistan Region, where the Peshmerga were offering stiff resistance. Baghdad itself was heavily fortified, with security forces, cement barriers, and armoured vehicles everywhere. Paul and I joined Minister Baird as he met with President Fuad Masum, Speaker Salim al-Jabouri of the Iraqi parliament, and Minister of Foreign Affairs Hoshyari Zebari. With each meeting, Baird reaffirmed Canada's support to Iraq in the fight against ISIS.

From Baghdad we flew to Erbil, staying overnight at the heavily guarded Divan Hotel. The flight to Erbil was in a Canadian Forces Hercules aircraft operating in the region. It was filled with military supplies destined for the Peshmerga. We wore flak jackets and helmets and sat in the darkened cargo hold during the flight. The next day, Paul and I joined Baird as he met with Peshmerga commanders and later, Masoud Barzani, president of the Kurdistan regional government. On the frontlines, Peshmerga soldiers pointed to the horizon to show us where ISIS fighters were dug in, about four miles in front of us.

We also visited schools serving as temporary refugee camps for displaced Iraqis who had fled from Mosul, and later we visited the new UNHCR refugee camp, Baharka, being set up so that Erbil's schools could be emptied of refugees, allowing teaching to resume. I sat down with a family that included small children and listened to the father describing the hardships they had faced and would continue to face until they could return to their home in Mosul. Before leaving, we also met briefly with Christian and Yezidi leaders.

Although we were in Iraq for only two days, it was long enough to understand the complex forces at play. Its Shiite majority and Sunni minority were often at odds, rarely united in a common purpose, except in this instance against ISIS, a ruthless sect of Jihadists hell-bent on destroying everything in its path as it tried to impose its extreme beliefs, not only in Syria and Iraq but ultimately beyond. Iraq was clearly under siege, and without military assistance from the coalition of countries against ISIS, which included Canada, it would certainly have fallen under ISIS control. Fortunately, its autonomous northern Kurdistan Region was also mounting a stiff resistance.

On a personal note, I was grateful to spend time with Paul Dewar and to get to know him better, away from the confrontational atmosphere of Ottawa. He held strong convictions, but he was also eminently reasonable. We both sat on the Standing Committee on Foreign Affairs and International Trade, and while he could have ignored my views

(I being the only Liberal on the committee), he was always receptive to any suggestion I made, allowing both our parties to occasionally adopt a common position to counter that of the majority Conservatives. Like everyone who knew him, I was very sad when he succumbed to cancer in 2019.

It was also a pleasure to get to know John Baird away from the confines of Parliament Hill. When I first met him in 2008, he had come across as ultra-partisan whenever he spoke in the House. He was also quick on his feet and often loud and boisterous. Observing him on this trip as he interacted with Iraqi officials, I saw a different John Baird and enjoyed some three-way, less guarded conversations with him and Paul. I got to know the real John Baird. (We would occasionally communicate with each other after he retired from politics.)

That fall, after almost seven years as an opposition MP, all of it under Prime Minister Stephen Harper, I reflected on whether the experience of politics had been a positive one for me. While seven years was much longer than I had ever expected, the answer was yes. Those seven years had allowed me to hone my skills as a parliamentarian while still contributing to my country. At the same time, I have to admit some days were tougher than others, particularly as the third party, when I wondered whether we would ever dig ourselves out of our hole.

I looked back at my run for the leadership and had no regrets. It had been a wonderful learning experience. I had also enjoyed my time as the critic for foreign affairs. Overall, I believe I had become an effective performer in the House, occasionally putting the government on its back foot, without resorting to rabid partisanship.

A few moments stick out in my mind. I had spoken on many occasions against the Conservatives' decision in 2012 to purchase the F-35 fighter aircraft when it was still in the development stage and experiencing a large number of problems. While many of these would eventually be resolved over a period of years, the F-35 was nowhere near ready for operational deployment when minister Peter MacKay announced the government's intention to buy it, and notwithstanding our Air Force generals pushing for the purchase, it was totally premature to select the aircraft. In fact, it was irresponsible. It also caused me to have serious doubts about the due diligence within the DND in promoting this choice, although Alan Williams, the Assistant Deputy Minister responsible for procurement at the time, did voice strong and legitimate concerns.

Most of the time, majority governments steamroll their way forward on the legislation they want to implement, but you can still be effective in opposition even when you don't get your way. Your job is to force the government to pay attention and think it through before acting. Sometimes that happens even though few people notice it.

A cause I championed, although unsuccessfully, was my effort to pressure the government to notify our allies that Canada would not join in any military operations if cluster munitions were to be employed. Canada had banned them and it didn't make sense to work with allies—e.g., the U.S.—if they planned to use them. I had learned about the devastation caused by these munitions, even decades later, when someone (usually farmers) accidentally stepped on unexploded ones. I had also lobbied John Baird to put Iran's Islamic Revolutionary Guard Corps, the IRGC, on Canada's terrorist list. I was not successful. Interestingly, the Conservatives would push for the same in 2023 with the Liberal government.

While I disagreed with many of Stephen Harper's positions, I did respect his skills as a leader and was grateful for the focus he put on our Armed Forces, particularly at the difficult time when some of our soldiers were coming home from Afghanistan in body bags. That respect had been long overdue. That said, you can only be positive about being in opposition for so long. I was eager to experience what it would be like to be in government.

Politicians are continuously tested. Each time you vote, you are making a decision, and the truth is that no one agrees with their party 100 per cent of the time. The same question arises before each vote: Will I toe the party line or, because of deeply held convictions, vote against my party? Every vote has implications. You may run afoul of your party, or you may run afoul of your constituents, but you should never run afoul of your conscience. The voters deserve that, and without it, trust in politicians disintegrates, and rightfully so.

One unique situation I had to deal with stemmed from being an astronaut in my previous career. There were those people, including in the media, who could not think of me in any other way, as though an astronaut could never be anything else in life except an astronaut. Sometimes this afforded me a measure of respect and at other times, the opposite, as in: "I can't believe this guy was an astronaut," or more commonly, "Garneau is still lost in space." I lost count of how many times people said that, thinking they were being funny. I just rolled with it. And, for the record, I always knew exactly where I was! Fortunately, the astronaut references became less frequent over time, and I was more often referred to as a member of Parliament, which is what I was.

As we rounded the corner into 2015, there was every expectation that after four years, Prime Minister Harper would call an election in the fall. The number of ridings would increase from 308 to 338 and some riding boundaries would change, including my own. I would be running in the new riding of Notre-Dame-de-Grâce–Westmount, losing Ville-Marie and gaining the western half of NDG and Montreal West. The latter was pretty solidly Liberal, and while Ville-Marie was more Liberal than the western half of NDG, I did not feel that the redrawing of my electoral boundaries would have a significant effect on my chances of being re-elected.

The election was called on August 4. The question on everybody's minds was whether Canadians were ready for change after almost a decade of Stephen Harper. After seven years in opposition, I certainly knew where I stood on the matter.

TWELVE

THE 2015 ELECTION would last for seventy-seven days as Stephen Harper attempted to win a fourth mandate. Tactically, he had decided on a long campaign, betting that the opposition parties would falter along the way. It was a big mistake, at least in the case of the Liberals. Although most polls showed the Conservatives and NDP coming out of the gate looking strong, with the remaining Liberals in third place, the picture had changed by mid-September, when it became a toss-up between all three parties. With a month to go, Justin Trudeau's performance in the televised debates and on the hustings went from strong to stronger. He came across as energetic, presenting fresh ideas that resonated with Canadians. The national momentum he generated was also trickling down to the riding level, benefiting my own campaign. Meanwhile, the NDP was beginning to free-fall and the Conservatives would flatline in the final two weeks as Liberal momentum continued to build.

On October 19, Canadians opted for change, with 39.5 per cent of them voting for a Liberal government. We had gone from a third party with 36 seats to a majority government with 184 seats. In my own case, I had secured 57 per cent of the votes in my newly redrawn riding.

Justin Trudeau deserved most of the credit for this historic turn-around. While fatigue with the Conservatives and Stephen Harper had been a factor, there was also undeniable enthusiasm for our leader. Also in our favour was the fallout from the public debate on the issue of the niqab. Thomas Mulcair had strongly defended a woman's right to wear the niqab at Canadian citizenship ceremonies, while Harper was pushing for a ban. Despite this principled stand, the NDP took a major hit in Quebec, which cost the party dearly.

The Liberals' time had finally come. There was a palpable air of excitement in our ranks. We would now get to steer the ship of state. I had waited a decade for this moment.

First on the agenda was the creation of a cabinet. A vetting process got underway, and I received a call telling me to report to a secret location. This was the first hint that I might be considered for a ministerial post. The meeting was to check for any skeletons in my closet. All the necessary questions were asked, and shortly after, I received a second call, this one telling me to report once again to the same location. This time, I would be meeting the prime minister–designate and learning my job assignment. Although I understood the reasons for secrecy, the process reminded me a little of *Mission: Impossible*, minus the self-destructing tape recorder.

While great care was taken to ensure I didn't bump into another colleague while waiting to meet the PM, I did spot one through an open door, but fortunately I knew how to keep a secret. As an ex-military, I knew the importance of maintaining the secrecy of classified information, and I would take the same approach throughout my time in politics and beyond. In my early years in opposition, I had repeatedly seen my party leak like a sieve, and I knew how destructive it could be. Nothing is more demoralizing than to see the content of a confidential caucus discussion laid out in detail by the media a half-hour later. As a new member, I had repeatedly pleaded with caucus to exercise discipline. On one occasion, based on clear evidence from a colleague, I confronted a suspected leaker, who vehemently denied my accusation—although, miraculously, the leaks stopped after that.

Following the vetting process, I naturally tried to guess what ministry I might be assigned. I thought of three possibilities, based on my résumé: Defence, because of my career in the navy; Industry, Science and Technology, because of my technical background, my astronaut career, and my time as president of the Canadian Space Agency; or Foreign Affairs, because Justin Trudeau had made me the critic for that portfolio during the two previous years. I would have been delighted with any of the three portfolios, although my first choice would have been Foreign Affairs.

Obviously, I was wrong. When the prime minister assigned me to Transport, I'm sure I looked surprised, because I certainly was. He came over to me, shook my hand warmly, and in the presence of Gerry Butts, his principal secretary, and Katie Telford, his chief of staff, told me the good news, no doubt expecting a big smile on my face. After picking up my jaw from the floor, I did my best to muster that smile. Rarely in my life had I been caught so off guard. Simply put, I had never considered Transport as even a slim possibility, although many friends jokingly told me afterwards that I was the natural choice because, after all, I had flown the space shuttle—which of course I hadn't (except in a simulator), but why disabuse them of the notion?

As it happened, it would take little time for me to fall in love with my portfolio, realizing it was not such a bad fit after all, given my previous experience. My career in the navy had exposed me to the world of marine transportation, and my experience as a private pilot, to the world of air transportation. Even my hundreds of weekly commutes on VIA Rail between Montreal and Ottawa had allowed me to understand the intricacies of rail travel.

All the new ministers were scheduled to be sworn in by the governor general on November 4. Two days before, as I left my home in Montreal, I took a spectacular five-foot plunge off the front steps, with my knees absorbing the impact as I hit the pavement.

Why was my first feeling one of relief that nobody had seen me fall? I was obviously far more vain than I realized. I blame it on my years as a politician. Notwithstanding the pain, I was able to stand and walk, or perhaps more precisely, limp. Despite the discomfort, I made my way on foot to my office, about a kilometre away, hoping the pain would subside. An hour later, when it became clear that it wouldn't, I hobbled home to find some medication and seek medical help. X-rays later that day would reveal damage to both knees, including cracked patellas.

My situation was not ideal, but after waiting ten years for this moment, I was damned if anything short of death would prevent me from going to Rideau Hall. Fortunately, Pam would be joining me for the occasion, an ideal support in every sense of the word. It dawned on me that if this had been a space shuttle launch, rather than being sworn in to cabinet, I would have been scratched from the event.

On the morning of the fourth, I met my future cabinet colleagues for the first time as we took a bus from downtown Ottawa to 24 Sussex Drive and then walked to Rideau Hall. I was using a cane, but I made it. Everyone remembers that glorious morning. The sun was shining and the fall foliage was at its peak. It felt like a new beginning. Obvious for all to see was the fact that, for the first time ever, the new cabinet was gender balanced. After all, it was 2015, said the prime minister.

That morning, I was sworn in to both cabinet and the Privy Council. What made a lasting impression on me was not so much the pomp and ceremony of the occasion, but the words of my oath—in particular: "I will in all things to be treated, debated, and resolved in Privy Council, faithfully, honestly and truly declare my mind and my opinion." I promised myself I would live up to that oath.

I could hardly wait to get going, beginning with high-level briefings from the ministry and the selection of my political staff. Ministers typically have about fifteen people, known as staffers, to assist them in their duties. Selecting the right people is critical, beginning with your chief of staff. That person is the primary conduit between you and the Prime Minister's Office, or PMO, and between you and your deputy minister's office.

In that position, I was fortunate to recruit Jean-Philippe Arseneau. His name had been given to me by someone in the PMO charged with trying to match up potential chiefs with ministers. Based on his CV, I decided to meet him for a coffee to see whether he might be a good fit. He hailed from the Magdalen Islands, had been deputy chief of staff to a senior Quebec minister, and wanted to broaden his experience in the federal arena. He had a calm demeanour and didn't seem like someone who could be easily rattled. Not being excitable was a prerequisite for me. He also had a sense of humour. We hit it off right away. After agreeing to be my chief, he began recruiting the rest of the team that would cover policy, communications, legislative matters, committee business, and the organizing of my work schedule.

Although I left it to him to choose the team, I did offer a couple of suggestions. One was Alain Berenstain, a friend and colleague with whom I had worked at the Canadian Space Agency. I thought he would make a good policy director because of his analytical mind. The other was Marc Roy, an experienced "comms person" who had advised me while in opposition and had worked in Prime Minister Martin's office.

It goes without saying that having good staffers makes all the difference. That said, political life is just as demanding on them as it is for politicians, and their employment is never secure. Most of the time, their fate is tied to yours. If you lose your job, they risk losing theirs. In addition, they work incredibly long hours, day and night, to ensure you're always kept up to date on the latest developments. But you can't expect them to stay with you forever. No matter how loyal they are to you—and mine were loyal—over time there is bound to be attrition. Some will decide to leave, and new ones will replace them.

For example, Jean-Philippe would step down as my chief of staff after three years to spend more time with his wife and new daughter, and Marc Roy would step in to replace him. Both Jean-Philippe and Marc were exceptionally loyal to me. I know this was sometimes difficult for them, especially if I took an approach that was at odds with the wishes of the Prime Minister's Office.

The first thing that struck me during my ministerial briefings was the economic importance of a well-run transportation network. As a trading nation, Canada's prosperity was directly tied to how well we moved our goods and commodities across the country and to the rest of the world. Our country is the second largest on Earth and subject to environmental extremes. This presents us with some daunting challenges. Dealing with them is critical. Which is why, first and foremost, I regarded Transport as an economic portfolio.

At the simplest level, transportation is about trains, ships, airplanes, trucks, and cars—anything that moves people and goods. Add to that the federal ports and airports and our vast railroad network, the entire airspace over Canada, and all our navigable waters and you begin to get a sense of the breadth of Transport Canada's responsibilities. It's a portfolio that touches everyone.

In addition to the travelling public, transportation is important to a large number of other stakeholders: our farmers, the mining industry, the forestry sector, our manufacturers, and all those who ship goods, not to mention those waiting for them. Then you have the environmental groups seeking to decarbonize transportation; Indigenous people, whose lives are often affected by transportation decisions; trade unions, who represent the tens of thousands of people who work in the transportation sector; and of course, the transporters themselves.

As someone who came to the job with prior knowledge of how ministries work, I realized the importance of the relationship I had to build with my department. While I might be responsible for giving it direction, I wanted communication to be a two-way street, which meant I had to be willing to listen to what my department had to say. For me, this was fundamental. Ministers who think their only responsibility is to tell their department what to do are not doing their job properly. Even worse, those ministers will push their department into being strictly reactive. Fortunately, my ministry and I believed in frequent and transparent communications, including regular Monday morning meetings that lasted three hours. This allowed me to stay abreast of the latest issues and, most importantly, to fully engage with my department. When I entered the House of Commons every Monday afternoon, I was up to speed and prepared to face whatever came my way, whether in Question Period, in committee, in House of Commons debates, or in media scrums.

Beyond my direct engagement with my department, I also needed to be in close contact with the transporters themselves, whether they carried people or goods. If they had complaints to make or suggestions to offer, it was important for me to hear directly from them to make sure they felt heard.

Finally, I needed to have an open-door policy so that my fellow MPs could bring transportation issues to my attention. Believe me when I say that every single riding in the country has transportation issues, some of them extensive and complicated.

Before my arrival at Transport, an independent review of the Transportation Act, in essence the ministry's bible, had been initiated by my predecessor, Lisa Raitt. The resulting report, authored by David Emerson and a team of experts, was made public just as I became minister. It was entitled *Pathways: Connecting Canada's Transportation System to the World.* The report contained a number of recommendations to make Canada's transportation system more efficient. It was a well-researched and comprehensive report, and I agreed with most of what it proposed. It would give birth to "Transportation 2030," my own strategic plan for the future of transportation, a plan that would drive most of what I undertook in the next five years.

I was fortunate to work with an excellent deputy minister, Michael Keenan, during most of my tenure. He in turn was supported by competent senior management. When I arrived, I made a point of meeting as many employees as possible, whether they worked in Ottawa or in our regional offices. They clearly loved their work and, as an engineer, I loved working with them. My background wasn't that different from theirs. I had been a civil servant before entering politics, albeit one in a military uniform and an astronaut flight suit. I understood them. Mostly, I recognized that although I might be the one telling them what to do, they had the more challenging task of making it happen, which any engineer can appreciate.

If you ask a minister to name their most precious resource, most of them will say time. You can't possibly meet everyone who wants to meet you, nor can you attend every event where your presence is requested. Hard decisions must be made, and this is where your political staff earns its keep. They will recommend how best to manage your time and, when necessary, cover for you. My schedule was carefully crafted, sometimes in ten-minute increments, with strategically placed breaks so I could prepare for my next meeting or simply catch my breath. It often reminded me of my days as an astronaut.

My staff called me "Buzz" among themselves and tracked all my movements. When I arrived somewhere they confirmed it by saying "The Eagle has landed," in obvious reference to the first Moon landing. I suggested "Pelican" rather than "Eagle" since I had enjoyed watching these fascinating birds up close during my time in Texas and Florida. And thus, it became "The Pelican has landed."

As I did my best to absorb the consistently incoming mountain of information, I began to zero in on the four main drivers of my job: the need to increase modern infrastructure in our trade corridors; to focus more on safety to minimize loss of life; to reduce GHGs and other pollutants from the transport sector; and, finally, to consult Indigenous communities, given transportation's influence on their lives. With these goals in mind, I was eager to begin.

THIRTEEN

AS A MINISTER, you face many challenges, including occasionally from your own colleagues. For example, pressure may come from the Prime Minister's Office or from caucus colleagues to do things a certain way, for reasons that are more political than anything else. Because I preferred to base my decisions on policy considerations, I would sometimes push back. Typically, I made decisions after a period of reflection, during which I listened to all the stakeholders, including my own colleagues, weighed all the facts and arguments, and even challenged my own department.

One decision I had to make, soon after becoming minister, concerned whether Billy Bishop airport, known to many as the Toronto Island Airport, should allow commercial jets. Porter Airlines, its major carrier, had expressed an interest in augmenting its fleet of turboprop aircraft with jets, including, possibly, Bombardier's new CSeries jet, known to be quieter and more fuel efficient than older jets of comparable size. A decision to allow jets had the potential of being good news for many people living in the Toronto area who found that flying from Billy Bishop was more convenient than Pearson International. That said, people living on the Toronto Islands, as well as some who lived nearby on Toronto's waterfront, opposed the use of commercial jets at Billy Bishop.

As with most decisions, there were other factors to consider. One was the Tripartite Agreement between the City of Toronto, Transport Canada, and the Toronto Port Authority (which ran the airport). The agreement stipulated that no commercial jets would be allowed unless the agreement was amended, which would require the unanimous consent of all three parties. Second, unbeknownst to me, the Liberal Party had promised our local candidate that commercial jets would not be allowed at Billy Bishop airport if we won the election. This fell solidly into the category of a decision made for purely political reasons.

When we did win, the newly elected Liberal MP began pushing for a public announcement that the government would not allow commercial jets at Billy Bishop. I found myself in a difficult position less than ten days into my portfolio. I did not yet have a political staff, nor had Transport Canada had a chance to examine the issue. I heard a range of opinions, both for and against. I suggested to the PMO that we should take the time to examine the matter, but was told in no uncertain terms to make a quick announcement. I complied, stating that Transport Canada would not be seeking an amendment to the Tripartite Agreement, thus effectively saying no to commercial jets. While this decision was well received by certain constituencies, it was certainly not by others.

I raise this issue to make a point. Decisions of this magnitude should be made only after carefully assessing all the factors that need to be considered, rather than being driven by a single one, which in this case was political. Whatever the reasons behind a decision, including political ones, governments must live with the consequences—hence my preference for basing decisions on thorough analysis. A decision that is supported by existing policy and scientific evidence can be defended.

What were the implications of allowing jets at Billy Bishop? To determine this, it would have been necessary to assess aircraft noise levels, operating hours, the need to extend runways, and expected passenger volumes and resulting changes to downtown traffic to and from the airport. Such an assessment by Transport Canada was prevented from happening. Quite possibly, it would have recommended against commercial jets, but without doing it, there was no way of knowing. Following the rushed Billy Bishop decision, I promised myself not to sign my name to any future decisions without the necessary due diligence.

In the case of Billy Bishop airport, the pressure came from the PMO. At other times, it could come from caucus. When members of my party brought issues to my attention, they were themselves under pressure from their constituents to get the minister of transport to "do something about it." Let me give you one recurring example: the problem of aircraft noise for those living under flight paths near major airports like Pearson, Trudeau, and YYC Calgary. Those affected wanted the flight paths changed or aircraft operating hours reduced, or both. At the same time, airports and airlines are expected to provide a high level of service to the public, consistent with safety and the regulations dictating how aircraft must approach and depart runways. Despite my best efforts, this was an issue where no matter what changes or improvements were made, it was impossible to satisfy everyone.

I mentioned earlier my concern about the major impact of transportation on the environment. In Canada, this impact is second only to that of the oil and gas sector when it comes to greenhouse gases, accounting for a quarter of Canada's total. Cars and trucks produce the vast majority of these gases, hence the urgency of transitioning to electric vehicles. My department was behind the decision to institute federal purchase incentives for EVs, a measure intended to get the ball rolling in changing people's mindsets and in accelerating the production of EVs by the major manufacturers. Some criticized this measure, but I believe it added important momentum to the transition. Quebec and British Columbia evidently felt the same way and had already taken the lead in offering incentives. The transition is beginning to happen, with more charging stations being installed and manufacturers ramping up production. The challenge, as with all new technologies, will be in balancing supply with demand, building more support infrastructure, and getting the cost down to reasonable levels.

Transportation produces other pollutants besides greenhouse gases. Carbon black expelled by ships is a concern in the Arctic. When these sooty deposits make the snow and ice surface "less white," it absorbs more solar radiation, causing more melting of the snow or ice cover. Also of concern are the discharge of wastewater from ships and the dumping of transportation waste products into the ground, some of which eventually pollute our aquifers.

Most Canadians are unaware that hundreds of abandoned vessels litter the Canadian shoreline. This situation led to a bill I introduced entitled "An Act respecting wrecks, abandoned, dilapidated or hazardous vessels and salvage operations." Its purpose was to address the issue of abandoned ships and small craft and to earmark funding to remove them from our coasts and inland waterways. Those still protruding from the water are unsightly, and those that have disappeared below the waterline are hazards to navigation and, in some cases, pollution hazards that can damage marine ecosystems, especially if their tanks still contain fuel.

I made a visit to Ladysmith Harbour on Vancouver Island, a veritable graveyard of abandoned vessels. Frankly, I was horrified. Dozens of small craft had been left there by their owners, some on the shoreline, some half-submerged, and some completely submerged. To describe them as eyesores is an understatement, as they littered an otherwise beautiful coastline, which, of course, didn't help with the tourist trade. Who knew whether they still contained fuel and other pollutants? Seeing it with my own eyes made me more resolute than ever to do something about it.

Beyond removing such wrecks, we also had to make owners accountable for end-of-life disposal of their vessels, instead of allowing them to simply walk away. In most cases, we were dealing with relatively small sailboats or motor craft, but in a few instances with much larger vessels, as in the case of the *Kathryn Spirit*, a 11,200 metric-tonnes bulk carrier built in 1967. For years, it remained beached on the banks of Lac Saint-Louis, on the St. Lawrence River near Beauharnois, a major eyesore for those living nearby.

As complaints grew about its unsightliness and the risk of pollution, I worked with Dominic LeBlanc, the minister of fisheries and oceans, to find a solution. This led to Transport Canada and the Coast Guard drawing up plans to dispose of the vessel. In December 2017, work got underway to patch and seal it, and then decontaminate it before taking it apart. Demolition was completed in October 2018. The good news is that a major hazard and eyesore was removed from the St. Lawrence River. The bad news is that owing to the insolvency of the foreign owner, the work came at considerable cost to the Canadian taxpayer. Just as mining companies must not be allowed to abandon their mining operations before cleaning up, we must ensure that boat and ship owners dispose of their vessels in the proper manner, or subject them to large fines and penalties. This was on my to-do list, but would require coordination with the provinces and territories.

One of the most important lessons I learned during my time at Transport was the impact many of our decisions had on Indigenous peoples. We've spent the past decidedly not bothering to consult them, despite the effect our actions have had on their way of life. One glaring example is building dams, with the resulting flooding of Indigenous ancestral lands. This had to change, and I was proud of the progress my ministry made. Tangible results achieved through Indigenous consultation included additional protections in the Canadian Navigable Waters Act, the implementation of the Oceans Protection Plan, and the Oil Tanker Moratorium Act, which prohibits oil tankers carrying more than 11,300 metric tonnes of crude oil as cargo from stopping, or unloading it at ports or marine installations along British Columbia's north coast. These were the result of listening, understanding, and taking action—something that had not been done before.

My department and I had worked hard with the Department of Fisheries and Oceans to create Canada's Oceans Protection Plan, which focused on protecting species such as the southern resident orca, responding to maritime emergencies such as ships in distress or pollution disasters at sea, and reducing the contaminants entering our waters from the land. I joined the prime minister in Vancouver to announce this important initiative. In designing the plan, consulting coastal First Nations was not only the right thing to do, it was the smart thing to do, because they had so much local knowledge and advice to offer. They had lived by and from the sea for millennia and were often the first out on the water when a vessel was in distress. We began harnessing that expertise and the willingness of First Nations to work together for a common purpose.

Working with the Indigenous community must become a natural reflex for government. We're making progress, but we're not there yet. Furthermore, it shouldn't be assumed that Indigenous groups will always agree with each other. Doing so is presumptuous and comes from a lack of knowledge of the Indigenous experience, past and present. Different communities sometimes see things differently, depending on their circumstances. One example was the west coast oil tanker moratorium. Some First Nations, remembering the *Exxon Valdez* oil spill, saw a potential ecological disaster in the making, fouling the pristine shoreline of Haida Gwaii or the Great Bear Rainforest, while others saw an economic opportunity in having a Pacific port other than Vancouver capable of shipping oil to foreign destinations. It was important to hear and examine both sides before making a decision.

One incident that clearly demonstrated the vulnerability of the west coast to oil spills was the sinking of the tug *Nathan E. Stewart,* which ran aground in October 2016 near Bella Bella in the Seaforth Channel on B.C.'s central coast because the crew member on watch fell asleep and missed a course change. As the tug sank, it released 110,000 litres of oil and diesel fuel into the Gale Pass, a significant marine harvesting ground for the Heiltsuk First Nation.

I flew to Bella Bella and met with Chief Marilyn Slett and other Heiltsuk elders. With an elder as my guide, I overflew the partially sunken tug and the area contaminated by the spill. This was clearly an ecological disaster for a community that has depended on its sea harvest for as long as it has lived in the Great Bear Rainforest. Although this was my first meeting with Chief Slett, I learned a great deal from her about the Heiltsuk people and their way of life; she was a strong, supportive voice when I presented Bill C-48, implementing the moratorium on oil tanker traffic along the west coast.

Of course, reconciliation is not simply about righting the wrongs of the past; it is also about providing opportunities for the future. That means jobs, including the kinds of jobs that have not traditionally attracted Indigenous people. To make the point, I'll use an example of an Indigenous-led initiative that inspired me. My colleague Mike Bossio had urged me to visit the First Nations Technical Institute's pilot training facility at the Tyendinaga (Mohawk) airfield in his riding. I was intrigued by the invitation, not only because of my interest in aviation, but also because it was an Indigenous-run school that was teaching Indigenous students to fly.

The school offers the only post-secondary Indigenous aviation program of its kind in Canada, providing hands-on flight training for students interested in pursuing a pilot's licence or working in the aviation industry. I was so taken with this that I made a visit and met with the people who ran it, as well as with some of the students and graduates of the program. I was more than impressed by this unique initiative, and I helped to initiate a grant from the federal government to upgrade infrastructure and aircraft. Most inspiring for me was meeting graduate women pilots who were flying with Canadian airlines.

When the time comes to discuss Justin Trudeau's legacy as prime minister, I personally believe that advancing reconciliation with Indigenous people may qualify as his greatest contribution. From my vantage point in cabinet, I was able to gauge his commitment, and while some of his proposed initiatives can justifiably be criticized, I did not doubt his sincerity. He began to change the tone of government in its dialogue with Indigenous people, from being condescending and paternalistic to being respectful and conciliatory, an ongoing process. Some will strongly disagree with me, arguing that his actions have not always followed his words and that his credibility has, on occasion, been undermined by bad judgment, but I believe Justin Trudeau deserves credit for moving the government more forcefully than ever before in the right direction. This new approach has permeated every government department, including mine, and although there remains much work to do, I give the prime minister high marks for the progress achieved so far.

A successful economy requires an efficient transportation system, and this can only exist if there are people to make it happen: the pilots and flight attendants; the aircraft mechanics; the people who run our airports; the air traffic controllers; the train conductors, engineers, and rail yard workers; the VIA Rail employees; the truck drivers; the ships' crews; the longshoremen in our ports; and so many more. If any of them go on strike or threaten to do so, a part of the transportation network slows down or grinds to a halt, with profound consequences for the economy. This happened on five occasions during my tenure: with longshoremen on the west coast and in the Port of Montreal, with rail workers at CP and CN, and with pilots at WestJet.

In each case, I found myself working with the minister of labour, at first Patty Hajdu and later Filomena Tassi, trying to avoid what could amount to costly work stoppages, while respecting the collective bargaining process. This usually required frank discussions with both sides, cutting to the chase and being honest about the issues at hand. I am proud to say that each situation was resolved without having to resort to back-to-work legislation, but it was rarely easy. Collective bargaining does work if given a chance. It is no coincidence that at such times, the public realizes just how much we all depend on the smooth operation of our transportation system. Without it, our economic underpinning is quickly brought to a standstill, which serves no one's purposes.

A good transportation system also requires modern infrastructure so that it can handle increasing volumes of goods and people and move them efficiently. Given Canada's vast physical distances and difficult climate, particularly in the North, this can be especially challenging. In some cases, it means having to build new infrastructure from scratch, and in others it means replacing existing infrastructure that has degraded over time. This was the purpose of the National Trade Corridors Fund, a multi-billion-dollar reserve focused on removing transportation bottlenecks and creating new capabilities, including in the North. Dozens of projects were funded through this program, with the federal government sharing the cost with the private sector and other levels of government. I was proud of my role in creating this fund, and I believe it was money well spent.

Not surprisingly, not every infrastructure project is well received. NIMBYism exists in Canada as it does everywhere, and a sensitive approach is required whenever new transportation infrastructure is proposed. On the one hand, my job was to maximize transportation efficiency. On the other hand, this could not happen by riding rough-shod over the concerns of those affected. CN's proposal to create a logistics hub in Milton for its freight operations in southern Ontario and beyond is a perfect example of a large infrastructure project that was important for the transportation of goods but that also impacted a community. In the end, the federal government gave the project the go-ahead, but only after CN agreed to certain conditions.

As a Montreal MP, I had spent years commuting weekly to Ottawa by train. I loved the experience and, over time, became quite familiar with VIA Rail operations. I knew every inch of the landscape from Gare Centrale in Montreal to the Ottawa train station. I also knew the state of the passenger cars, most of which were rapidly reaching the end of their useful life. It was time to replace them and their locomotives. Transport Canada initiated the replacement program and, after a competition, VIA Rail awarded the contract to Siemens. The new cars and locomotives, built in Sacramento, California, began arriving in 2022.

Most people who take the train will tell you they enjoy the experience. It's more relaxing and worry-free than air travel, although frequency of service and punctuality are issues. As for cost, it is less expensive than flying and can be competitive with single-passenger car travel if you take into account the cost of fuel and the wear and tear on the vehicle, not to mention insurance and registration costs.

Shortly after my arrival at Transport, I received a briefing about a VIA proposal called High Frequency Rail, or HFR (not to be confused with the much more expensive high-speed rail, which uses bullet trains). The High Frequency Rail concept, as the name implies, means more frequent trains in VIA's busiest corridor, between Quebec City and Windsor, Ontario, which carries over 90 per cent of its passengers. The concept was premised on the theory that if trains ran more often, passenger volumes would increase. A crucial element of the proposal was that VIA would run on its own dedicated tracks instead of having to use freight train tracks, where it is required to give right of way to freight, resulting in delays. Dedicated tracks would allow greater speeds and more on-time arrivals. Finally, HFR trains could be electrified over most of their route, a plus for the environment.

Over time, the HFR concept evolved to exclude the Toronto-to-Windsor segment because the Ontario government had initiated its own feasibility study of high-speed rail in that corridor. It would now be an HFR corridor between Quebec City and Toronto, with new routing through Peterborough in Ontario and Trois-Rivières in Quebec. Determining where the track lines run is a critical consideration, since it triggers environmental assessments, consultations with First Nations, and the possible expropriation of land. While the HFR project is more complex than many people realize, it holds the promise of high-quality, cleaner, and faster passenger rail travel in Canada's busiest corridor.

I championed HFR from the beginning, although it was not entirely clear to me where the prime minister stood. This ambiguity led me to request a rare face-to-face meeting with him to argue for the project, but even then, I came away somewhat unsure of his position. I realize that prime ministers have to juggle many priorities at the same time, but my sense was that, while Trudeau saw the potential merits, he was not yet sold on a project that would take a decade to complete. My job, therefore, was to continue building the case. Despite a lack of direction or decision-making from the top, my department and I persevered for four years, finally securing enough funding to undertake initial development. Much heavy lifting was required, but I believe I got it over the hump. Recently, as I left government, I was happy to see that the initial procurement process was getting underway. Needless to say, I hope HFR becomes a reality.

Sometimes governments make the wrong decision, with heavy consequences. One such mistake, made under Pierre Trudeau, was the expropriation of agricultural lands in the Mirabel area, north of Montreal, to build an airport. This occurred in the late 1960s and affected roughly three thousand families. The project was intended to turn Mirabel Airport into the main international airline hub for eastern Canada. But that vision never came to be, due in part to the absence of rapid transit from Montreal. The passenger terminal was demolished in 2014 and the airport is now used primarily for cargo flights.

Close to 38,800 hectares of some of the best farmland in Quebec had been expropriated to make way for the project. Of these, 32,000 hectares that had remained unused were sold back in the 1980s, in some cases to the farmers who had originally owned the land and were now renting it. In 2006, under the Harper government, the remaining lands outside the airport zone were also sold. However, the government excluded a wooded area of 300 hectares that was inaccessible by road. That obstacle was removed in 2016 when Ottawa and the City of Mirabel reached an agreement to provide access to this last bit of territory.

On April 15, 2019, fifty years after the decision to expropriate, I went to Mirabel city hall and announced that the government of Canada was selling back the last remaining parcel of land. I told the group of assembled citizens, most of whom had lost their lands, that we had made a big mistake and that we were sorry for what had been done to them. While this did not erase the mistake, for many in attendance, it was a cathartic moment, a moment of closure.

FOURTEEN

MOST OF THE TIME, the public pays little attention to what goes on in the world of transportation until it affects them directly, such as delays at the airport, noisy trains near their home, or the closure of a major road. The exception to this is when an accident occurs. When that happens, the public wants answers right away, and the spotlight focuses squarely on Transport Canada. The first and most relevant question to be asked is, Did we do everything possible to prevent this from happening?

One of my main responsibilities was to minimize the risk of accidents. Consequently, safety was always on my mind. As transport minister, there was never a moment when I could completely relax. I lived with the constant awareness that I might at any time get a phone call telling me that something had gone horribly wrong on the roads, on the rails, on the water, or in the air. My department and I had to be prepared for that possibility.

As much as we strive to make transportation safe, accidents happen, and when they do, there can be victims and loved ones left to grieve. In 2014, I witnessed that grief first-hand when I met the citizens of Lac-Mégantic, the town in southern Quebec where forty-seven people died on July 6, 2013, when a runaway train loaded with crude oil careened into their downtown area at more than a hundred kilometres an hour, derailed, and exploded in a massive inferno. I had been asked by my party to represent them at a church service to commemorate the first anniversary of the tragedy. The church was full, and I stood outside with many citizens as we listened to moving testimonies about sons and daughters, brothers and sisters, fathers and mothers lost forever. Everyone in the town knew someone who had died that night. This horrific event had traumatized the residents of a close-knit community of five thousand, some of whom are still experiencing PTSD to this day. At the time, I had no idea I would become Transport minister barely a year later.

In 2016, when I was in the post, I was invited to Lac-Mégantic for a townhall, to meet the citizens and to listen to them. Hundreds turned up, and not surprisingly, it was an emotional gathering. While some vented their anger, others wanted to know what the federal government was going to do. Many spoke of the need to divert the train track so that it would no longer pass through their downtown. One person stood up and held a railway spike in his hand for all to see, a spike he had recovered from the track. He looked my way and said, in front of everyone, that he was giving it to me as a symbol, to remind me of my responsibilities as minister of transport. More specifically, it was to be a reminder of the forty-seven people killed in his town, and of the need to do better. I was tremendously moved by what this man said, and I put the spike on my office desk the next day, where it remained for the duration of my tenure. I often picked it up in the years that followed and just held it, reminding myself that people's lives depended on how well I did my job.

I would return to Lac-Mégantic in May 2018 with the prime minister and Quebec premier Philippe Couillard for the announcement of a twelve-kilometre rail bypass project that both governments had agreed to fund. I had worked hard to make this happen. When I addressed the assembled crowd, I took the spike out of my pocket and held it high to remind everyone of its personal significance to me. Following the announcement, I had the opportunity, with the prime minister, to meet with a small group of citizens who had lost loved ones. They were still grieving; people's lives had been changed forever.

When I left Transport, I passed the Lac-Mégantic spike on to my successor, Omar Alghabra, who also kept it on his desk. It is my hope that this spike will be passed on to future Transport ministers, regardless of party, as a stark reminder that in transportation, lives are always at stake.

Sadly, it was not the last time I would meet with the families of loved ones who had died in tragic circumstances. Accidents occur in all modes of transportation. On April 6, 2018, a bus carrying the Humboldt Broncos junior hockey team hit a semi-trailer truck on the highway near Armley, Saskatchewan, after the semi-trailer had failed to yield at a flashing stop sign. One cannot imagine a more devastating event in the life of a small, tightly knit community.

This tragedy led to two important changes: Transport Canada made the installation of seat belts in coach buses built from 2020 onwards compulsory, and Alberta, Saskatchewan, and Manitoba undertook to review the minimum entry-level training requirements for truck drivers.

Air travel is considered one of the safer ways to get from one place to another, but no mode of transportation is without risk. As a private pilot, I had made a point of reading aircraft incident reports to avoid potential problems during my own flights. Many of the incidents I read about involved small planes and were rarely publicized, while others received national or even international attention. Two former federal ministers, Jean Lapierre, once a transport minister himself, and Jim Prentice, who had been premier of Alberta, both tragically lost their lives in private aircraft accidents in 2016, while I was at Transport.

On March 10, 2019, a major tragedy garnered worldwide attention: the crash of Ethiopian Airlines Flight 302, claiming 157 lives, including 18 Canadians. The aircraft was a Boeing 737 Max 8, flying out of Addis Ababa. This was the second crash of a Max 8 in less than five months; Lion Air Flight 610 had plunged into the Java Sea on October 29, 2018.

Following the Lion Air crash, an international investigation determined that the aircraft had pitched down repeatedly as the pilot fought to bring the nose back up, in what would turn out to be a losing battle. Attention focused on a little-known software module, the Maneuvering Characteristics Augmentation System, or MCAS, which might have triggered the command to push the aircraft's nose down.

The U.S. Federal Aviation Authority was the global lead in the safety certification of the Max 8, as it was designed and manufactured in the U.S. After the Lion Air crash, the FAA issued a notice, called an airworthiness directive, informing airline operators around the world about this issue with the MCAS system, but indicating that the plane remained safe to operate. At the time, Air Canada, WestJet, and Sunwing owned a total of forty-one Max 8s. No country grounded its aircraft pending the results of the investigation. However, as a precautionary measure, Transport Canada required that all the Canadian airlines flying the Max 8 modify the pilot response instruction card in case something similar happened on one of their aircraft, effectively instructing the pilot to perform steps that would disable the MCAS. We were the only country in the world to do so.

When Ethiopian Airlines Flight 302 went down on March 10, many countries grounded their Max 8s as a precautionary measure without knowing the cause of the crash or citing a specific reason. The approach we took in Canada was different. The aviation safety officials at Transport Canada scrambled to pull together information on the crash. I pressed hard for analysis as quickly as possible, to have some evidence to support a safety decision in response to the crash. That evidence came from NAV Canada in the early hours of March 13 in the form of information made available by Aireon, a company whose technology, automatic dependent surveillance-broadcast, or ADS-B, determines an aircraft's position in three dimensions via satellite navigation or other sensors and broadcasts it, allowing it to be tracked.

Within a few hours of obtaining this data, the aviation safety experts in the department concluded that the flight profiles of the Lion Air and Ethiopian Air flights were remarkably similar, indicating some unknown common factor behind the crashes. Based on this evidence, the experts recommended grounding the Max 8 as a precaution. I concurred and gave the direction to move quickly.

For me, as an engineer, it was the evidence supporting the safety determination that made the difference. We would be the first country in the world to ground the Max 8, based on our careful analysis of data. I held a press conference that same morning to explain our decision and presented the evidence supporting it. Just before I made this announcement, my chief of staff, Marc Roy, placed a call to Todd Inman, chief of staff to Elaine Chao, U.S. secretary of transportation, to inform her of Canada's forthcoming grounding of the aircraft and the reasons for it, including the evidence from the Aireon data. Transport Canada aviation experts made similar calls to their counterparts in the FAA. Later the same day, the U.S. took the same action, citing the same evidence, grounding the Max 8 effective immediately.

At that point, we had no inkling about the facts that would come to light in the investigations regarding the grossly negligent and irresponsible behaviour at Boeing related to this new MCAS system, which led the FAA to unknowingly certify a plane with such a serious safety flaw. I was shocked, the experts in my department were shocked, and Canadians were shocked.

It would be revealed that Boeing had not provided information about the MCAS system to be included in pilot manuals or for training purposes and, in its rush to get the Max 8 certified, had downplayed its significance when communicating with the FAA, the agency responsible for oversight of the new aircraft type.

As an astronaut, the majority of my training had dealt with the "what ifs": how to recognize a wide variety of malfunctions and then respond. Every conceivable failure scenario was considered, resulting in the development of fully tested procedures to deal with it. Boeing had clearly not fully tested how the MCAS software would respond in the case of a single angle-of-attack sensor feeding it faulty data.

We vowed the Max 8 would not be allowed to fly again in Canada until we were certain the safety problems that had claimed 346 lives had been completely fixed. This required leadership on our part: Transport Canada officials worked closely with their counterparts in Europe and Brazil to carefully review the work of the FAA on the required safety changes to the Max 8, and insisted they include a re-examination of the overall certification of the plane, a full audit of the fixes to MCAS, full flight testing for the plane (including by Transport Canada pilots), and additional simulator training for all pilots who would eventually fly it again, a requirement that I had publicly stated was necessary.

The actions on which we insisted represented a shift in Canada's relationship with the FAA. For decades, Canada had accepted the FAA certification of new American aircraft types to be flown in Canada, without Transport Canada undertaking a full certification of its own. Given the FAA's professional reputation, a simpler process of validating the FAA certification was considered sufficient. This had worked well for decades and avoided expensive, time-consuming duplication. Because of Transport Canada's equally professional reputation, the FAA had taken a similar approach to aircraft and aircraft parts designed and built in Canada and first certified by Transport Canada. For example, they had accepted Transport Canada's certification of the CSeries commercial jet models designed and built by Bombardier (now known as the Airbus A220). However, the results of the Max 8 investigation pointed to the need for Canada to re-evaluate how it validated FAA-certified aircraft, beyond the unprecedented measures we took in relation to the Max 8.

In late 2019, I met for three hours with some of the families of the eighteen Canadian victims of Ethiopian Airlines Flight 302. As was the case in Lac-Mégantic, lives had been shattered by the loss of loved ones. We sat together around a boardroom table, and each of the families spoke in turn. Anger and frustration were expressed, sometimes intensely, by those who felt there had not been enough accountability for what had happened. Some argued that all countries should have grounded their Max 8 aircraft after the Lion Air crash, something that no country did. (While Canada had modified its pilot procedures after the Lion Air crash—the only country to do so—I knew this would not change the way the families felt.) I could only listen, offer my condolences, and reiterate the support measures we had put in place for the families. I felt powerless to comfort them, and that made me feel awful.

I learned an important lesson that day. While Transport Canada was focused on what had happened to Ethiopian Airlines Flight 302, I personally should have been quicker to reach out to the families who had suffered loss. Sadly, this lesson would be put to the test less than a year later when we responded to another air disaster, that of Ukraine International Airlines Flight 752, shot down by Iran as it took off from Tehran. This time, I would reach out from the start.

For me, every transportation accident was a moment of reckoning. In addition to any actions my own ministry might take, the Transportation Safety Board, an independent watchdog, could also choose to investigate and then issue a report of its own, with recommendations to advance safety, some of which would be aimed at Transport Canada. They also maintained a Watchlist, updated annually, of what they considered the ten most urgent safety concerns to be addressed. We regularly reviewed this list. My long-term goal was to make it as small as possible.

Every transportation accident gave rise to questions: What caused the accident? Could we have prevented it? Did we fail to address a risk through additional safety measures? Was it human error? Was it an equipment malfunction? Did we approve a design that wasn't safe? Did we miss something during an inspection? Do we need to make changes? More than anything else, transportation safety was the concern that kept me awake at night, that was constantly on my mind throughout my tenure as minister.

The goal of safety is to minimize risk. In the ideal world, no risk is acceptable. Unfortunately, this is unrealistic. If that were a requirement, nothing would move. It's also important to remember that it's not just about equipment safety standards and regulations; it's also about the training of vehicle operators, about their working conditions, about the maintenance of equipment, and about predicting situations such as dangerous weather. Human error, equipment failure, and unexpected events such as avalanches and other "acts of God" can also play a role in accidents. The truth is that many things we do involve a degree of risk. I, like everyone else, had to deal with this reality every day as a minister and, more significantly, had grappled with it when I chose to fly to space. That said, we should always strive to make transportation as safe as possible and do so proactively.

A case in point is the state of the pilots who fly us to our destinations. It goes without saying that we expect them to be sober, and of course there are rules and regulations to address this. We also expect them to be fresh and alert, which raises the question of pilot fatigue: How many hours should a pilot be allowed to fly before the risk of an accident caused by fatigue becomes unacceptable? We all know of car accidents where the driver fell asleep at the wheel. Some of us have pulled over to the side of the road when we felt that overwhelming urge to close our eyes. The same can happen to a pilot, although the option to pull over isn't available.

It was my duty to ensure that pilot fatigue never contributed to a transportation accident. Transport Canada periodically updated pilot "duty day" regulations, based on the latest scientific evidence about fatigue. I raise this issue because my fellow caucus members expressed strong opinions about it, some of which were at odds with what my department and I were planning to do. This is a good illustration of the challenge of making the right decision when others may be pushing back.

On the number of hours a pilot can fly safely before reaching his or her limit, factors other than total flight time must be considered. One is when the pilot's duty day begins. Is it a daytime flight or an overnight flight? (We are all familiar with circadian rhythm.) Another factor is how much rest the pilot requires before their duty day begins. Also to be considered are the cumulative hours of flying in a given time frame—for example, the number of hours per month. These and other possible determinants must be taken into account. They could result in an airline's decision to carry an extra pilot on long overnight flights. Finally, what happens in the case of emergencies where pilots might have to exceed their flight time limits, such as during medical evacuations or while fighting forest fires?

While Transport Canada was guided by the latest science, organizations such as the Air Transport Association of Canada and various pilot associations wanted other factors to be considered and, to that end, lobbied Liberal members of Parliament, who then lobbied me. Some argued that the regulations my department and I were proposing were more restrictive than required and would harm business; others argued that they would still lead to unacceptable risk. Although not explicitly stated, two of the reasons behind the airline sector's lobbying efforts were related to a shortage of pilots, particularly in the North, and the cost of having additional pilots on the payroll.

Notwithstanding my colleagues' personal interventions, I was unwavering in my efforts to implement the new regulations. I was even summoned before Treasury Board officials to argue my case, an unusual step. I stuck to my guns because I believed it was the right thing to do. Safety had to trump all other considerations. In the end, I was able to move forward with the changes I had proposed. The bottom line is that decisions at Transport were sometimes difficult owing to the number of stakeholders involved and did not always receive unanimous support. That did not excuse me from having to make such decisions and to be fully accountable for them.

Let me provide another example, one we are all familiar with. From time to time, Transport Canada updated the list of items prohibited on aircraft to ensure it included objects or substances that posed a threat to crew or flight safety. For a time, all knife-type blades were considered a risk, including the metal knives provided for in-flight meal service. Of greatest concern was protecting the flight crew from direct attack, which led to measures such as reinforced and locked cockpit doors on commercial aircraft.

In due course, a decision was made to end the ban on knife blades shorter than six centimetres on domestic flights. Such blades were no longer considered a credible terrorist threat. Yes, they could injure someone (as could other objects allowed on flights), but they were not considered dangerous enough to bring down an aircraft, given all the new measures in place. Unfortunately, the decision became politicized.

In the fall of 2017, Transport Canada's decision came up in the Quebec legislature, which passed a near-unanimous motion demanding, on the grounds of safety, that the government maintain its ban on all knives on commercial flights, no matter their size. This occurred in part because the Bloc Québécois had been arguing in the House of Commons that the decision to allow small-bladed knives was for no other reason than to accommodate religious Sikhs wanting to wear kirpans. It is worth noting that a unanimous motion, also on the grounds of safety, had been passed by Quebec's National Assembly in 2011, banning Sikhs wearing kirpans from entering the legislature. The Quebec government frequently adopts such motions as a means of exerting pressure on the federal government, even though their motions carry no force in law. The willingness of the Liberal Party of Quebec to go along with these motions clearly illustrates the fact that federal Liberals and their Quebec counterparts are, in many respects, two different species of Liberal.

Understandably, this caused a great deal of concern in the Sikh community, given the kirpan's religious significance. The truth was that this should never have been about kirpans. It was about knife blades in the context of flight safety. Blades shorter than six centimetres had been approved, and this would also apply to kirpans, whether worn or stowed in carry-on bags. Nevertheless, the decision created a stir in Quebec after the Bloc Québécois sounded the alarm in the House of Commons and the media. Quebec's most popular current affairs and entertainment program, *Tout le monde en parle*, invited me on as a guest—an opportunity for me to explain the decision we had taken or, if I failed to do so convincingly, look bad in front of a lot of people. It was not uncommon for the hosts of the show to be aggressive with their political guests. The stakes were high.

My chief of staff, Jean-Philippe Arseneau, and my director of communications, Marc Roy, understanding the importance of this event, grilled me with every possible question I might be asked. I was glad they did, but as it turned out, the hosts were not aggressive and they accepted the arguments I presented. In fact, they invited me to stay for the rest of the show while other guests were interviewed, and I was even invited to ask them questions. As it happens, one of the guests had written a book debunking well-known conspiracy theories, including one that theorized that the Moon landings had all been staged in a film studio. Needless to say, I thoroughly enjoyed helping him disprove one of the most persistent myths about the Apollo program.

Incidentally, in the years since, there has not been a single incident of a passenger threatening a flight with an approved short-bladed knife or kirpan. While no decision is completely risk-free, I believe we made the right one in this case.

For the most part, my focus on safety was to minimize the death or injury of humans, but sometimes the victims were animals. We have all seen roadkill on our highways. We hear about aircraft hitting birds and trains hitting forest animals. Ships sometimes run into whales and other marine mammals that are disoriented by engine noise and unable to get out of the way. In the past, we mostly shrugged when such incidents occurred, our main concern being humans.

Times have changed. The death of North Atlantic right whales from ship collisions and entanglement in fishing gear in the Gulf of St. Lawrence became a pressing concern for me and the minister of fisheries and oceans in 2017, when twelve died in Canadian waters. This endangered species, now estimated to number around 350, was migrating farther north in search of copepods, the tiny crustaceans on which it feeds, as the copepods moved into the colder waters of the gulf.

The government needed to act quickly, and it did. The fisheries minister and I mobilized our departments, and within weeks put measures in place. Fisheries and Oceans tackled the problem of fishing gear entanglement, and Transport Canada announced restrictions on vessel speed in designated sea lanes within the gulf. We also monitored traffic and levied fines on non-compliant ships. All this required prompt adjustments by both fishers and marine vessel captains, but they accepted the changes as necessary. Over time, our measures were refined and the number of right whale deaths in Canadian waters was dramatically reduced.

As we implement measures to protect endangered species such as the North Atlantic right whale, the belugas of the Saguenay estuary, and the southern resident orcas of the west coast, we must fully acknowledge that transportation can be a factor in the death of these mammals and make every effort to minimize that risk. In other words, the powerful imperative of fast and efficient transportation must sometimes yield to an even greater imperative, that of preserving endangered species. It will be many years before we know whether we have succeeded or failed in this regard.

When it comes to safety, there will never be a quiet moment when nothing is happening. New challenges will keep appearing— for instance, the problem of distracted driving, or people flying their recreational drones into controlled airspace or pointing lasers into the cockpits of aircraft on final approach. If you imagine these to be minor safety issues, think of what may happen when fully autonomous vehicles deploy in large numbers, speeding down our highways and through our neighbourhoods, controlled by an array of sensors and the latest artificial intelligence. If we don't do our homework to ensure this happens safely, we will pay a price.

FIFTEEN

IN MY CAPACITY AS TRANSPORT MINISTER, I frequently interacted with caucus members but had little interaction with the prime minister, other than during cabinet meetings. I believe a few of my colleagues were close to Justin Trudeau—for example, his childhood friend Dominic LeBlanc, whose opinion he valued—but I was not. Although we shared Liberal values, we didn't have a great deal else in common, which is fine. It's certainly not a prerequisite of the job. During my five years at Transport, I only remember him calling me twice to see how I was doing. It was pro forma, and he did not seek my opinion on any issues. Generally speaking, he was not overly interested in Transport, unless a problem surfaced.

For example, it was customary before annual G7 leaders' meetings for a number of ministerial pre-meetings to take place. The hosting country would decide which ministers should meet. Transport ministers met when Japan was the host in 2016 and when Italy was the host in 2017. However, when it was Canada's turn in 2018, the PM dropped Transport from the list, despite my efforts to have it included, given the international nature of transportation and the many challenges we shared with other G7 countries. I found this decision regrettable, to say the least.

My approach to engaging the PM one-on-one was to do so only if I felt it to be absolutely necessary. I did so on two occasions during my five years at Transport: I met him face-to-face in his office to advocate for High Frequency Rail and I talked to him on the phone to discuss the financial situation of the airlines at the height of the COVID pandemic.

That being said, I certainly spoke up at cabinet meetings if I had something significant to say, but I also prided myself on being a low-maintenance minister who got on with his job and avoided making a show of it. That's how I had worked all my professional life. If I thought of saying something, I asked myself: Will this contribute to the conversation? I didn't feel the need to hear my own voice. My NASA experience as a capcom in Mission Control had shaped me in this respect. Be clear and be succinct. Otherwise, just listen. My experience as a military officer had also shaped my approach, especially when it came to the PM. I viewed him as my commanding officer, and although I didn't always agree with him, he was the boss.

For a caucus, team building is extremely important; hence the two-hour Wednesday morning meetings when we gathered to discuss the issues of the day, hear new ideas, and allow members to vent, a needed catharsis for some. In addition to these weekly meetings, a couple of times a year we attended three-day retreats so that we could take deeper dives into policy matters and spend more time team building in a less formal setting. The same would happen at the regional caucus level. This allowed the Quebec caucus, for example, to focus on issues of importance in that province.

We did the same in cabinet. In addition to our weekly three-hour meetings, we held retreats to discuss future policy and build team cohesion. Although most of our meetings were stimulating and productive, there were those, especially in the early days, that focused on the management theory known as "deliverology," which is essentially a method of ensuring timely delivery on government promises, which sounds perfectly reasonable, except that it's so obvious as to be totally unnecessary to discuss. Quite frankly, as an adult entrusted with cabinet responsibilities, I found it silly. Although I went along with it to be a team player, I was happy when the subject faded completely from our lexicon and was never mentioned again.

In addition to my portfolio at Transport, I was called upon to assume other responsibilities. Over time, I was a member of the following cabinet committees: the Defence Procurement Committee, the Indigenous Reconciliation Committee, the Ad Hoc Committee on the COVID Response (technically I was an alternate member, but was asked to attend every meeting—a no-brainer given the critical impact of COVID on transportation), the Global Affairs and Public Security Committee, and, briefly, the Intelligence and Emergency Management Committee (once the PM or someone in PMO realized there was an important element of transportation in almost every emergency). This was a valuable opportunity for me to broaden my knowledge of other government portfolios.

At the outset of the committee on our response to COVID, we knew almost nothing about the virus. Like every other country on the planet, we were groping in the dark, scrambling to respond in the most intelligent manner. It was like being at war without knowing your enemy, particularly in the early days when so many were dying. All sorts of potential responses were discussed, with some being adopted, others rejected. Should we try to close off the country and isolate ourselves or should we institute strict border controls, including proof of a negative COVID test result within the past forty-eight hours? Should we make everyone wear masks, and if so, what kinds of masks other than N95s were considered safe? How could we ensure that those most in need of personal protective equipment were prioritized, given the shortages at the outset? How could we help keep our hospitals from being overwhelmed by patients requiring ventilators and in some cases being put into induced comas?

There was of course no vaccine at the outset, and we didn't know how long it would take to develop and approve one. Our only guidance to Canadians was based on previous virus outbreaks. We were told that COVID was most likely spread by expelled droplets from an infected person, droplets that could travel up to two metres. When I raised the possibility with Health Canada of COVID being transmitted through ventilation systems (by aerosol propagation), I was told there was no evidence to support this. That's how little we knew at the time. Later, as we gained greater knowledge, that position would change.

Fortunately, as pharmaceutical companies scrambled to develop a vaccine, a promising breakthrough in the delivery of messenger RNA, or mRNA, occurred, allowing vaccines to be developed in less than a year, a remarkable accomplishment. Also fortunate, Minister of Procurement Anita Anand and her department placed orders with a number of the most promising vaccine suppliers, increasing our chances of receiving early deliveries.

Unlike other cabinet committees in which I participated, the Ad Hoc Committee on COVID was unique in that it dealt with an ongoing crisis where we were having to make many decisions, often in the absence of complete information, hope for the best, and adjust as new facts came to light. In hindsight, I believe the committee did its job in a competent manner.

Back in 2017, while NAFTA 2.0 negotiations were underway, I had the pleasure of chairing the Cabinet Committee on Canada–U.S. Relations and being part of the "Team Canada" effort to convince U.S. federal and state politicians of the importance of trade with Canada. This included my participation on a three-person panel in Washington, D.C., with Larry Kudlow (President Trump's economic advisor at the time) and Jesús Seade (Mexico's deputy minister of foreign affairs), at the annual meeting of U.S. state governors. I spoke about the serious impediment posed by U.S. plans to impose tariffs on steel and aluminum, stating that time was running out and that this could be a deal-breaker. I was very direct in pointing out that, for many states, Canada was their number one export destination, and for many of those same states, Canadian imports played a major role in their economy.

During the time I chaired this committee, I occasionally met with Kelly Craft, the U.S. ambassador to Canada. While most of our interactions occurred during official meetings, I also got to know her and her husband, Joe, on a more personal level and we became friends. They were proud Kentuckians, and in 2018 they invited Pam and me to join them in Louisville for the Kentucky Derby. Whether or not you're a fan of horse racing, when someone invites you to the Kentucky Derby, you go. We flew down, covering our own travel and hotel expenses. Like everyone who goes to the Derby, we dressed for the occasion. Pam wore a spectacular hat, the first time I had ever seen her wearing anything on her head other than military headgear or a toque.

We joined Kelly and Joe for dinner the night before, where I had the opportunity to speak to my American counterpart, Secretary of Transportation Elaine Chao, also a Kentuckian. She attended the Derby the next day, joined by her husband, Senator Mitch McConnell (yet another Kentuckian). Pam and I received the royal treatment, including a special tour of Churchill Downs and its inner workings. We even placed some bets and came out ahead, mostly due to Pam's inspired choices.

I was touched by the warm friendship and hospitality the Crafts extended to us. It had not been an easy assignment for her, given Canadian sentiment about Donald Trump, the man who had appointed her. She would go on to become the U.S. ambassador to the United Nations after her time in Ottawa.

Earlier, I wrote of going to the White House in 1984, when I was an astronaut, and meeting President Reagan and Prime Minister Mulroney in the Oval Office. It had been a memorable occasion in my life, but not one, I thought, likely to be repeated. However, a second opportunity did present itself when, in February 2017, Prime Minister Trudeau made his first official visit to Washington to meet the incoming president, Donald Trump. He took several ministers and staff with him, and I spent about half an hour with Vice President Mike Pence in the company of a few colleagues, as well as U.S.

generals Jim Mattis and John Kelly. I found Pence to be pleasant, although deeply conservative. I mostly enjoyed meeting the two generals (both straight shooters) and would later have dealings with General Kelly when he led Homeland Security before his brief tenure as Trump's chief of staff.

The meeting with the VP was followed by lunch with the president. About twenty of us, Americans and Canadians, sat at a large dining table with the PM opposite the president. Various subjects were discussed, but at one point Trump spoke about the advanced technology of the U.S. Air Force's F-22 Raptor stealth tactical fighter. He pointed out that this aircraft could accelerate vertically. In response, Trudeau pointed to me and said that I had flown on the space shuttle. Trump looked at me, sized me up for a couple of seconds, and said something to the effect that I'd aged well. I thanked him for the compliment, and that was my short conversation with Donald Trump. I suppose I fared better than many.

Although I had hoped my incredible mother would somehow go on forever, she died on May 7, 2017, at the age of ninety-four. The week before, she was eating a cheeseburger at our house and nothing seemed amiss. Three days later, she was having dinner with my brother Charles and, again, everything appeared fine. The following morning, however, she felt unwell and was hospitalized. Three days later, with most of us by her bedside, she died peacefully from what the doctor called organ failure. It was that fast. Her body had decided to shut down for good.

While her passing was sad for me and all those who knew her, I'm grateful she did not have to live through the COVID pandemic, given the forced isolation and the death of several people in her retirement residence. She really did live an extraordinary life, and all who knew her remarked on her sharp mind. She was amazingly healthy until the end. I will always miss being able to pop over to see her and have a chat.

I mentioned earlier that becoming a grandfather was a moment of transition in my life, a joyous one to be sure. The death of my mother also represented a transition, although this time, a sad one. The generation preceding me was now gone. I no longer had a living parent, and my older brother, Braun, was also gone. I was now the oldest in our family.

Later that summer, Pam, Adrien, George, and I joined Yves and Simone and their families in Provence, in the village of Bonnieux in the Luberon hills. I had always dreamed of visiting Provence, and finally doing so with my entire family was an incredibly special time I will always be grateful for. We toasted my mother on numerous occasions, wishing she could have been with us. She would have loved it.

Governments and their leaders are prone to falter over time. As mistakes are made, the shine wears off. Some mistakes are forgiven, others not. The difference often has to do with the size and scope of the error and, perhaps most important, how it is handled.

A case in point occurred in early 2019, and it would eventually lead to ministers Jody Wilson-Raybould and Jane Philpott leaving politics. The SNC-Lavalin affair, as it was referred to, concerned the government interfering with a criminal proceeding involving high-level corruption and fraud. It should never have happened, and it wouldn't have if Prime Minister Trudeau and others in the PMO had not tried to influence Attorney General and Minister of Justice Wilson-Raybould, something no one, including—and perhaps most especially—the prime minister, is allowed to do. The Ethics Commissioner would eventually rule that the Prime Minister had contravened the Conflict-of-Interest Act by using his position of authority to influence Wilson-Raybould in an attempt to get her to overrule the decision by her director of public prosecutions not to negotiate a deferred prosecution agreement (DPA) with SNC-Lavalin. Under a DPA, charges can be stayed and ultimately dropped if the offending party fulfils all DPA conditions.

This was a serious error in judgment, made worse by obfuscation, outright denial, and the demonstration of little or no remorse on the part of the prime minister, which promptly led to Treasury Board president Philpott resigning in protest as a matter of principle. Sadly, there is a tendency over time for those at the top to think they can control everything. In this case, the prime minister had clearly overstepped his authority.

While I personally would have preferred to see a deferred prosecution agreement with SNC-Lavalin, that decision rested solely with the attorney general and her director of public prosecutions. Sadly, we would lose two senior ministers and the fallout would damage the party and its brand, ultimately leading to the departure of Gerry Butts, the PM's closest advisor. While as Liberals we all bask in the glory of our PM's achievements, we also take a hit when he makes a serious mistake, and I was certainly one of many that were unhappy with this turn of events.

After being justice minister and attorney general, Wilson-Raybould was appointed minister of veterans affairs, but resigned soon after. Following her decision to leave, the PM called me into his office and asked me whether I would agree to switch ministries and take over at Veterans Affairs. He left me the option of saying no. At the time, I had important initiatives I wanted to complete at Transport, but because I was also an ex-military, I offered to take the new portfolio provided I could keep my current ministry. Without explaining why, the prime minister said he was not open to this. I therefore declined his request and continued at Transport. In retrospect, I wished this meeting had never taken place. I felt like a pawn in a political game. I also felt that moving me to another portfolio showed a lack of respect for the work I was doing at Transport.

Elections are always moments of reckoning: they're when you receive your report card. Canadians would go to the polls on October 21, 2019, and although I fared well, winning 56 per cent of the votes in my riding, the government was reduced to minority status, going from

177 to 157 seats. Our share of the popular vote had dropped by more than 6 per cent, from 39.5 per cent to 33.1 per cent. In fact, more Canadians had voted Conservative than Liberal.

With SNC-Lavalin still fresh on everyone's mind, the prime minister and his brand had clearly taken a hit, although not enough for Canadians to want to replace him with someone else. In contrast to the genuine enthusiasm that had greeted him in 2015, the response this time was definitely more subdued.

As often happens after one mandate, voters take note of the mistakes committed, but, barring a complete disaster, they allow the incumbents a second chance. Fortunately for us, the Conservatives under Andrew Scheer did not present a strong enough alternative to govern, even though they increased their seat count from 95 to 121. The Bloc Québécois also increased its count, going from 10 seats to 32. They had clearly risen from the ashes, invigorated under their new leader, Yves-François Blanchet. While this did not signify a rise in separatist sentiment, it did signal once again that Quebecers vote strategically, always looking for the best deal. As for the NDP, they had tumbled from 39 seats to 24, as Jagmeet Singh contested his first election as the party's leader.

As we would soon discover in so many ways, governing with a minority was far more challenging than having a majority.

SIXTEEN

I HAD ABSOLUTELY NO IDEA WHAT TO EXPECT, especially after turning down the prime minister's request that I go to Veterans Affairs, so I was pleased when I was reappointed minister of transport. There was still a lot I wanted to do, and this would be the perfect way for me to end my political career, having agreed with my family that I would not run again. They had been dropping hints that they would be pleased to have me around the house more often. I had turned seventy and I thought it was a good idea, while I was still healthy, to plan a departure point, preferring not to make a decision on the spur of the moment. I also believed that retirement required mental preparation, it being a major transition in one's life. Because we were now a minority government, I estimated that the next election was roughly two years away, sufficient time for me to complete some important work before announcing that I would not run again.

Although no one could anticipate what lay ahead, the next two years would turn out to be a very challenging time for Canada, including in transportation. In quick succession, I would have to deal with four important, back-to-back events, beginning with a Teamsters strike at CN, no less.

Our railways move staggering volumes of goods across the country and to the ports that ship our commodities and products to the world. Most Canadians don't fully appreciate the critical role they play in our economy. We become impatient with trains moving through our neighbourhoods, making noise and slowing traffic. We complain about late deliveries, no matter what the cause. Sadly, people are sometimes killed at crossings and sometimes trains derail. I'm the first to say that our railways must do better when it comes to safety and being attuned to community concerns, but I also recognize the important work they do.

While our railways routinely deal with landslides, avalanches, flooding, and forest fires, they usually recover quickly. A strike, on the other hand, raises the possibility of a prolonged interruption. On November 19, 2019, the Teamsters union walked off the job at CN. I worked closely with the outgoing labour minister, Patty Hajdu, and her successor, Filomena Tassi, to see how we could bring everyone together to find a resolution. I also met separately with each side, as the strike exposed a particularly serious problem with the delivery of propane to Quebec. Propane is used by farmers to dry their grain and to heat their barns, and 85 per cent of it came by train from Sarnia refineries. It would be impossible to fill the pressing demand by other means. Wet grain spoils—hence the urgency of the situation. Quebec ministers were already breathing down my neck.

Interacting with both sides is always delicate. While encouraging them to bargain in good faith and as quickly as possible, I could not insert myself into the negotiations, much less suggest any specific fixes, and I certainly could not take sides. That said, the Teamsters and CN care about their reputations and what Canadians will think of them if the trains are stopped for too long. Fortunately, both sides came to an agreement without arbitration, and after eight days, CN resumed operations. We all breathed a sigh of relief.

Sadly, tragedy would strike early in the new year, triggered by the assassination of Gen. Qasem Soleimani, head of the Iranian Quds Force, considered a terrorist organization by both Canada and the U.S. Following his death in an American drone strike near Baghdad airport, Iran swore revenge, firing missiles at U.S. forces stationed in Iraq. My deputy minister called me in Montreal on January 7 to inform me that, as a precautionary measure, Transport Canada had advised Air Canada not to overfly the area where missile firings might occur. That same day, we received the news that an aircraft had crashed shortly after takeoff from Tehran airport.

Throughout the night, details began emerging, and for Canada the news was nothing less than a national tragedy: Ukraine International Airlines Flight 752 had crashed with 176 people onboard, including 55 Canadians and 30 permanent residents. I immediately drove to Ottawa to meet with my department.

Overnight, Transport Canada examined the ADS-B position data from the flight, showing its trajectory as it took off and climbed to altitude. The data indicated a normal takeoff and climb before the signal dropped out abruptly when the aircraft was passing through 8,000 feet. The next morning, during an incident response meeting convened by the prime minister, I informed him of this and joined him at a press conference where I reiterated our findings to the media—adding that while it was too early to speculate about the cause, the data suggested that "something very unusual happened."

As the government continued to investigate, it became clear that the only credible explanation for such a major malfunction was either a bomb going off onboard or a missile strike. In the hours that followed, multiple intelligence sources confirmed a missile strike. The prime minister made this public the following day and reiterated the need for a full investigation.

At first, the Iranians denied this, but as conclusive evidence mounted, they could no longer maintain the lie, and three days later they admitted shooting down the aircraft, claiming it was an accident. So many promising lives had been cut short and the Iranian Canadian community was in a deep state of shock. They had lost spouses, children, parents, and friends.

Canada demanded that Iran provide a full and transparent account of what had happened. Foreign Affairs Minister François-Philippe Champagne met with his counterparts from Sweden, Afghanistan, the U.K., and Ukraine, countries that had also lost citizens. He and I would meet with the families of victims in the months following, to listen to them and to assure them that Canada would not relent in its efforts to get to the truth and to make Iran accountable.

Even getting the aircraft's voice and data recorders analyzed required a monumental effort. After speaking to my French counterpart and receiving confirmation that France was willing to examine the black box data—one of only a handful of countries with the expertise to do this—I engaged in repeated exchanges with Iranian officials to get them to send the boxes to Paris as quickly as possible, something that should have occurred in a matter of days but that instead took several months. Simply put, Iran obfuscated at every turn, refusing to cooperate, much less be open and transparent. Everything smacked of an attempt to cover up the truth. Frankly, I was not surprised, given the track record of this despotic regime that was a major backer of terrorist organizations in the region and a threat to global security. They were clearly doing everything possible to avoid complete accountability for killing 176 innocent people. Instead of investigating the entire chain of command, Iran was clearly intent on shielding those in the higher echelons of the Islamic Revolutionary Guard Corps, the armed force that defends the regime.

It's important to note that the skies over Tehran were filled with commercial aircraft before and after Flight 752 was shot down, even while Iran expected retaliatory strikes and had set up defensive surface-to-air missile batteries following their own attacks on Iraq. In other words, Iran continued to allow aircraft to take off and land at Tehran airport, and no effort was made to prevent flights from entering Iranian airspace—either a decision that was deliberate and utterly reprehensible or a striking demonstration of incompetence, with unfortunately tragic consequences.

Following the tragic loss of Malaysia Airlines Flight 17, shot down by the Russians over Ukraine in July 2014, the Netherlands attempted to get the international community to institute a warning system that would prevent airlines from flying through conflict zones. Yet nothing had been put in place. After the loss of Flight 752, Transport Canada followed up with a similar initiative, called Safer Skies, which we presented to the International Civil Aviation Organization. I appeared before the ICAO council on two occasions to advance the proposal, which is now under development.

A few weeks after the downing of Flight 752, several First Nations began erecting rail blockades across the country, to protest the construction of the Coastal Gaslink pipeline through 190 kilometres of Wet'suwet'en First Nation territory in B.C. These acts of civil disobedience, affecting both freight and passenger operations, presented federal and provincial governments with a difficult challenge: how to respect the right of Indigenous people to protest while keeping the trains moving. The railways filed multiple injunctions, including against the Mohawk near Belleville, but the provincial police chose not to act on them, clearly concerned that their actions might provoke violence. As a result, the railways had no choice but to divert their trains, which in one instance included CP allowing CN to use its tracks.

At the federal level, it became clear that the problem would persist, and possibly escalate, unless the government intervened and met with the Wet'suwet'en hereditary chiefs to try to find a solution, a step undertaken by Minister Carolyn Bennett in late February.

Meanwhile, dark clouds were gathering on the horizon, and in early March, Canada's chief public health officer, Theresa Tam, advised against large gatherings as part of the country's response to the rapidly growing COVID-19 pandemic. Ironically, the emergence of a major global health challenge contributed to bringing the blockades down. By the second week of March, most were gone.

Just like the CN strike, the blockades had managed to seriously disrupt freight and passenger transportation. I understood why they had been set up, given their effectiveness as a way to protest, yet I found it frustrating that our economy had been taken hostage. Personally, I do not believe that we should allow blockading of Canada's rail transportation network, including on Indigenous land, given the danger to life and the major economic impact blockades have on Canadians.

With the brutal onset of the pandemic, there would be no time to catch our breath. Throughout 2020 and onwards, we would deal with a host of COVID-related issues, including the repatriation of tens of thousands of Canadians stranded abroad. This was a major undertaking, requiring Foreign Affairs Minister Champagne to negotiate landing rights for Canadian airlines in several countries that had closed their airports to foreign carriers and requiring my department to work with airlines to bring the stranded Canadians home.

As the pandemic took hold, it also became apparent that cruise ships had become highly contagious incubators for the virus, and many passengers were not allowed to disembark at ports-of-call such as Vancouver, Quebec City, and Sydney, Nova Scotia, until they could be tested and, if necessary, isolated. Like most countries, Canada would implement a temporary ban on cruise ships operating in its waters.

We acted quickly on social distancing and making masks compulsory when travelling by air, by train, or on ferries. On rare occasions, special measures were instituted to temporarily ban aircraft from entering Canada from certain countries with high rates of COVID. Proof of vaccination became necessary for passengers using federally regulated transportation. Preflight testing also became compulsory for those coming to Canada, as well as testing and quarantine upon arrival. For a time, body temperature measurements were used for screening passengers before flights.

Finally, subsidizing the cost of passenger and cargo flights to Canada's North and other remote regions became necessary, given that small northern carriers were a vital part of the supply chain and had lost most of their passenger revenue.

———

One issue that provoked a strong negative backlash was that of cancelled flights. When Canadian air carriers suffered a precipitous drop of 90 per cent in passenger traffic during the pandemic, they had to deal with thousands of flight cancellations and opted to offer credits for future flights instead of ticket refunds. They argued that they were not obliged to reimburse customers since the cause of the cancellations was beyond their control. In addition, no one knew how long the pandemic restrictions would last, adding uncertainty to the recovery.

The public's anger was totally justified, exposing a shortcoming in the air passenger bill of rights that I had introduced the year before. No one had anticipated the case of a pandemic causing massive cancellations over a lengthy period of time. In December 2020, I instructed the Canadian Transportation Agency (CTA) to address this shortcoming. New reimbursement requirements in the case of cancellations beyond the airlines' control would come into effect in September 2022.

The idea of a charter of air passenger rights had been around for some time, having been debated in the House (but not adopted) when I was an opposition MP. I personally found the arguments in favour to be compelling. Other countries had implemented passenger rights, and I felt it was time for Canada to do the same. Their purpose was to hold airlines accountable for flight cancellations and delays within their control. The airlines, without exception, were not happy with this and typically invoked safety considerations or the fact that delays could occur for other reasons, such as baggage handling or security issues, keeping aircraft at the gate.

The key, of course, is defining "within their control," particularly as that relates to safety, such as when an equipment failure occurs right before departure. Should passengers be compensated in such a case if the failure results in a significant delay, especially if it means a passenger will miss a crucial event? The airlines argue that it would be dangerous for passengers if pressure were put on airlines to cut corners when it comes to safety.

Another problem was the backlog in processing complaints. When the charter was first announced, the CTA had anticipated an increase in the number of complaints, necessitating hiring more people to process the higher volume. Accordingly, I asked for additional funding for that purpose. I recognize that finance ministers must make difficult choices when it comes to deciding what to fund, but there are also consequences—in this case a large backlog of complaints, not to mention frustrated air travellers.

Without a doubt, it is challenging to create a charter of air passenger rights that both sides deem to be fair. Nonetheless, it is the government's responsibility to find the most equitable solution.

While the health-related measures put in place during the pandemic were difficult for passengers, they were also difficult for those who work in the transportation sector. It was critical to continue moving goods, whether across the country, across the Canada–U.S. border, or around the world. This presented serious challenges, given that personal protective equipment was initially in short supply and vaccines had yet to be developed. We owe a great deal to our transportation workers who maintained operations at our airports, our ports, and our railway yards, not to mention our truckers, who continued to move goods when most rest stops were closed. All ensured that vital goods such as food, medicines, and other essential products reached Canadians.

Canada, like every other country on Earth, endured a difficult 2020 as it hunkered down. The hope of quickly developing an effective vaccine was the light at the end of the tunnel that sustained us all. Fortunately, the government preordered the most promising vaccines under development, ensuring that we could quickly begin to administer them as soon as they became available. In mid-December, the first small batch began to arrive, and most Canadians were ready to roll up their sleeves. Gradually, the situation improved as more and more people were vaccinated and rates of mortality decreased.

As 2021 dawned, I felt that the transportation sector was up to speed on measures to cope with the pandemic. Yes, new variants were possible (as would be the case with the delta and omicron variants), but we now knew what to do to ride out the storm. While measures imposed on travellers remained onerous, they were necessary until the situation improved to the point where we could relax the rules on PPE, testing, and proof of vaccination. Meanwhile, people and goods were moving.

SEVENTEEN

AS 2021 DAWNED, I was certainly not expecting a move from Transport, given that we were not yet out of the woods with COVID-19. However, the prime minister had other plans. In a mini cabinet shuffle in January, he sent me to Foreign Affairs.

Only he knows why he made this decision. What is clear is that the appointment was triggered by the departure of my colleague Navdeep Bains, who, for family reasons, had decided to leave politics. At the time, he was minister of industry, science and economic development, a key portfolio. With his departure, the PM decided that the best person to fill the position was François-Philippe Champagne, a capable minister who was then minister of foreign affairs. Hence the reason for my move, with Omar Alghabra replacing me at Transport.

During a tenure that had lasted more than five years, I had often paused at the wall outside my office, where photographs of all the transport ministers since 1936 were displayed, with a blank spot reserved for me the day I left. After Lionel Chevrier, who served for nine years, and David Collenette for six and a half years, I was the third longest serving transport minister, which gave me a quiet sense of pride. Unquestionably, you get more done the longer you stay in the same job. And I wasn't feeling burnt out by it: yes, the last fifteen months had been particularly challenging, but they had also been a time when the ministry stepped up to the plate and made a positive difference in the lives of Canadians.

Of course, nothing lasts forever. When I left, I did so with a heavy heart because it meant saying goodbye to the great people with whom I had worked for so long. It meant letting go of projects I had initiated and that were still underway. While I looked forward to my new responsibilities, I knew I would be looking over my shoulder occasionally to see how "my" transportation projects were progressing.

Cabinet shuffles occur quickly, and, as an experienced minister, I was expected to hit the ground running. Because I needed to get abreast of a whole set of new files, I didn't have the luxury of time to return to Transport Canada to pack my boxes and say goodbye. COVID didn't help either, with so many people working from home. In addition to receiving briefings from my new ministry, I also needed to assemble a new team of staffers, since Transport and Foreign Affairs were two very different portfolios.

After shuffles, the majority of your existing staffers stay behind to assist your successor, and you inherit some of the staffers already in place at your new ministry, assuming they are willing to stay on to work with you. That way, incoming ministers can get up to speed quickly with people who already know the files. I was fortunate to inherit some excellent people, which helped greatly with my learning curve.

My shuffle into Foreign Affairs raised questions about how important this portfolio was in the mind of the prime minister. I would be the fourth incumbent in a little over five years. On the one hand, I was overjoyed to be assigned this challenging responsibility, one that I had always hoped to take on one day. On the other hand, it definitely bothered me that the turnover rate was so high, suggesting that the prime minister did not view it seriously enough, perhaps because he felt he was really in charge of it. In my opinion, Foreign Affairs needed long-term stable leadership, and I was hoping I would be the one to provide it.

This of course begged the question of how much freedom I would be given to make decisions. I had enjoyed considerable latitude when

I was at Transport, but would the PM and Deputy PM Chrystia Freeland be more hands-on when it came to Foreign Affairs? Based on what I knew, I had to think they would.

I was sworn in to my new portfolio via a virtual meeting with Rideau Hall. The PM had called me the previous week to announce his decision and I was certainly pleased. Having been the critic for the portfolio when Liberals were in opposition, and also having sat on various committees dealing with international issues, I felt ready for the job.

Most Canadians probably assume that a prime minister will occasionally summon a minister for consultation or to have a private one-on-one conversation about a particularly serious or pressing matter. While I had hoped this might happen when I took over at Foreign Affairs, a senior ministry, it did not. My new responsibilities did not bring us closer. In fact, I was called only once and that was to discuss the status of the two Michaels imprisoned in China (more on this later), and it was in a boardroom with various other people from the PMO and the Privy Council Office, in preparation for a meeting with our ambassador to China, Dominic Barton. The prime minister's aloofness led me to conclude that he did not consider my advice useful enough to want to hear from me directly, relying instead on his staff. I found this disappointing, to say the least. The expectation was that communications between him and me would be via the PMO, and so consequently I never knew what information, if any, reached him.

Like many, I believe too much power and control are centralized in the Prime Minister's Office. It's been said a thousand times before, but that doesn't make it any less true. Although Justin Trudeau would counter this idea with statements like "my door is always open," the PMO in fact isolates most ministers from the PM, even though ministers are directly accountable to him through their mandate letters. There may be an element of "good intentions" in doing this, so as to protect the PM's precious time, but power and control also play into it, and it's obvious that Justin Trudeau either welcomes this or tacitly approves. Ultimately, the responsibility for correcting this unacceptable concentration of power rests with the prime minister. He promised voters and everyone in government that he would do this when he took over in 2015, but he did not. Meanwhile, I got on with my job.

Foreign Affairs is unique in that Canada's relations with other countries are based not only on the policies we adopt and the alliances we create, but also on the personal relationships that the PM and the internationally focused ministers develop with their counterparts. That requires a sustained, long-term effort. It's important for a minister of foreign affairs to remain in his or her portfolio for a reasonable amount of time, preferably several years. Being the fourth appointee in a little over five years sent a different message. Our allies could logically question whether Canada attached sufficient importance to this portfolio, and they did. On virtually every one of my introductory calls to my counterparts, I heard the following polite comment: "I hope you will be in the post longer than your predecessors"—a not so subtle message. That was certainly my hope.

Obviously, Foreign Affairs is important for a host of reasons. Our economy is critically dependent on trade with other countries, our national security requires a collective approach with our allies, our ability to mitigate climate change and other global challenges requires international cooperation, and, finally, Canada has always aspired to be a voice for democracy, human rights, and the international rules-based order.

With that in mind, I have often asked myself how the average Canadian sees it. My impression is that Canadians are generally more focused on domestic issues than international ones. That said, we are a country of immigrants, and some feel strongly about our relations with certain countries, perhaps because they or their ancestors came from one of them and still have ties to it. Canadians also feel strongly about protecting democracy, with many having fled less democratic countries. What I'm also confident in saying is that Canadians care about our reputation abroad. They care whether Canada is viewed favourably by the rest of the world. They like to think that we are admired and that our voice matters. Who can blame us for thinking such things?

Unfortunately, Canada's standing in the world has slipped, in part because our pronouncements are not always matched by a capacity to act or by actions that clearly demonstrate that we mean what we say. We are losing credibility.

I believe Justin Trudeau has overestimated Canada's impact abroad. The PM's missions to China in 2016 and 2017 and to India in 2018 were not successful. None of them lived up to expectations. We were not properly prepared. At a fundamental level, we did not understand who we were meeting. We thought we could seduce and were surprised it didn't turn out that way. Gone was the clear-eyed approach of a prime minister like Jean Chrétien, who always knew with whom he was dealing and who forged pragmatic alliances with world powers.

We live in a world that is constantly changing, a world that requires a sustained long-term focus to ensure we really understand it. It is not sufficient to pay attention only when a concern arises, something this government has made a habit of. We must pay attention all the time. In other words, Foreign Affairs must be proactive, not reactive. The expression that Canada punches above its weight hasn't been credible for a long time, and some no longer believe that Canada is carrying its weight as a democratic middle power. We can correct this, but it will take work, time, and a consistent message that is followed by action. That was certainly my objective as I began my new job.

My first month at Foreign Affairs involved briefings from ministry officials and many calls to other countries. Although the portfolio would normally involve travel, the pandemic had grounded everyone, and we conducted our diplomacy over the phone or by videoconference. High on my agenda were our challenging relations with China on many fronts, including that nation's treatment of Uyghurs in Xinjiang and its arbitrary detention of two Canadians, Michael Kovrig and Michael Spavor. Both had been charged with spying and been imprisoned soon after Canada had agreed to an American request to detain Meng Wanzhou, the chief financial officer of Huawei, in Vancouver pending her extradition to face charges in the United States.

Also on my radar, and a file I knew well from my time at Transport, was Iran's refusal to make itself fully accountable for shooting down Ukraine International Airlines Flight 752.

In addition to Afghanistan, which I will address further on, I would also deal with other issues: helping to build our relationship with the Biden Administration after four years of Donald Trump; Russia's continued aggression towards Ukraine; the resurgence of violence between Palestinians and Israelis; the military coup in Myanmar; the continued exodus of citizens from Venezuela under the Maduro regime; the threat by Michigan to close the Enbridge Line 5 pipeline supplying fuel to Ontario and Quebec refineries; the assassination of Haitian president Jovenel Moïse; and the war in Tigray, Ethiopia. Just another day at the office!

On February 15, I hosted an international videoconference in which Canada and fifty-seven other countries announced that they had endorsed the Declaration on Arbitrary Detention in State-to-State relations. (As of this writing, seventy-five nations have signed.) This Canadian initiative, originated by my predecessor François-Philippe Champagne, stated that a country had no right to detain innocent citizens of another country to exert pressure on that country. Arbitrary detention had certainly occurred in the case of the two Michaels, but it was not unique to China, even though the Xi government took the declaration personally and pushed back aggressively. That was China's standard reaction whenever it was criticized.

I should point out that my views on how China approached foreign relations were partly coloured by my experience with their officials during my time at Transport. At the time, China had been pushing for greater access to specific Canadian airports for their subsidized, state-owned airlines but was not prepared to agree to Canada's demands for reciprocal concessions at specific Chinese airports, including the new airport in Beijing. This would put our airlines at a disadvantage in the lucrative tourism business developing between Canada and China before COVID-19. Although many of my cabinet colleagues disagreed with me at the time, wanting to promote tourism from China at seemingly any cost, I pushed back. I never understood the reasoning for giving China, or any other country, the upper hand in such negotiations. It was my job to ensure fairness for Canada's transportation sector. Sure, there could be some advantages for Canada, but getting bullied and caving in to their one-sided demands seemed to me a bad strategy, especially in the long run.

On February 22, the Conservative Party presented a motion in the House of Commons stating that China was committing genocide against Uyghurs and other Turkic Muslims. The motion passed unanimously. I was the only cabinet minister who voted, abstaining on behalf of the government. While there was strong documented evidence of China's mistreatment of Uyghurs in Xinjiang province, the government's position was that China should allow United Nations human rights experts immediate and unfettered access to Xinjiang to fully examine the allegations of mistreatment. And let me emphasize the word *unfettered.*

Fifteen months later, China would finally authorize a visit to Xinjiang by Michelle Bachelet, the UN high commissioner for human rights, although her access to potential evidence could not be described as unfettered. She released her report three months later, and while she does not use the word *genocide*, she leaves little doubt that China is guilty of international crimes, including possible crimes against humanity.

Predictably, China fought hard to block the report and responded with the usual denials and outrage. Whether its treatment of Uyghurs constitutes genocide in accordance with the Genocide Convention, or is a deliberate state-sanctioned plan to assimilate them, which would effectively be "cultural genocide," one thing is undeniable: at a minimum, China is guilty of international crimes, based on its deliberate actions to persecute, marginalize, and assimilate Uyghurs and other Turkic Muslims.

The prime minister held his first meeting with President Biden on February 23. It was a virtual meeting, and I joined Chrystia Freeland and the PM in his office. This was an opportunity to reset the clock on Canada–U.S. relations. Several topics were discussed, including working together to defeat COVID-19, maintaining strong trade between our two countries, fighting climate change, fighting racism, strengthening North American defence, and securing the release of the two Michaels, who were still, after two years, languishing in Chinese prison cells. The atmosphere was relaxed and it was refreshing to hear President Biden's views on so many topics of mutual interest. After four years of Donald Trump, it was clear that we were once again dealing with a president who shared long-established values with Canada. Suffice it to say that climate change, fighting racism, and working together to defeat COVID would never have been on the agenda with Biden's predecessor. At one point, the PM happened to mention that it was my birthday, and the president spontaneously broke into a shortened version of "Happy Birthday."

By then, I had already spoken with my counterpart, Secretary of State Antony Blinken. He had called me an hour after being confirmed in his position, which I took as a good sign. I knew he spoke French because he had attended the Lycée in Paris in his youth. I told him of my own schooling at the Lycée in London and that immediately connected us. In fact, we broke into French for a while. Right from the beginning, Blinken's extensive experience in state and national security matters stood out. He was a consummate professional, as he would ably demonstrate in the ensuing months and years. He possessed a calm, low-key approach and a pleasant demeanour. I liked him from the beginning and really enjoyed our relationship as we interacted on a number of issues during my tenure.

Because of our challenges with an increasingly powerful and assertive China, I focused a good deal of effort on our China strategy—work begun under my predecessor. Over and above the country's human rights record and the arbitrary detention of the two Michaels, there were a host of concerns: its inappropriate use of trade sanctions, its tightening grip on Hong Kong, its treatment of Tibetans and other minorities, its growing presence in the South China Sea, and its aggressive posture towards Taiwan. However, China was also a trading partner, and we needed to work with it and other countries on global challenges such as climate change. It was essential for us to remain engaged, although not at any cost.

Let me make a point about the China strategy being developed when I arrived: it was not influenced by the detention of the two Michaels. Their detention was simply one of many realities with which we were dealing. It had not prevented us from announcing the Declaration on Arbitrary Detention.

The China strategy was central to my mandate. I was focused on getting it approved and getting the government to make it public. As I've said before, foreign policy should be proactive. We owe it to Canadians to let them know where we stand on China, not to mention the importance of letting China itself know. In my opinion, the strategy was ready for rollout in the spring of 2021. However, despite my best efforts, I did not get the green light to present it to cabinet. Bearing in mind that his trips to China had not gone well, I believe the prime minister and his entourage were hesitant to put the strategy out, in part because the two Michaels were still detained. I think that was a mistake, pure and simple.

Notwithstanding this silence, I could not avoid questions about China coming my way and I began to speak publicly about "the 4Cs" and how they would define our relationship. We would coexist, cooperate, compete, and, when necessary, criticize without mincing words. As for the latter, we had already repeatedly condemned China for its crackdown in Hong Kong and had imposed sanctions on four Chinese officials and one state-run organization over their treatment of Uyghurs. It was also crystal clear that China detested our Declaration on Arbitrary Detention and the growing number of countries signing on to it.

Meanwhile, I was working on another strategy being developed by the department, our Indo-Pacific strategy, and was making progress, but I was once again unable to get it to cabinet, let alone to the Global Affairs and Public Security Committee, which had to approve it before it went to cabinet. Given the number of issues that the cabinet sees, getting anything on its weekly agenda, or before that on the agenda of a cabinet committee, requires a triage by the PMO and PCO (which takes its direction from the PMO), with priority items obviously going first. My sense of urgency for an Indo-Pacific strategy was obviously not shared.

The Indo-Pacific is now the most dynamic economic region in the world, but it is also a region beset by tensions with potential global repercussions, in particular the aforementioned aggressive posture China is taking towards Taiwan and in the South China Sea, and North Korea's continued expansion of its nuclear weapons program. It was more important than ever for Canada to increase its engagement in this region and to develop a long-term strategy, for both economic and security reasons. We had begun such an engagement on the economic front back in the 1990s under Chrétien but had not sustained it, let alone built on it after he stepped down. As a Pacific country, Canada had a great deal to learn about our Asian neighbours, including the countries known as the ASEAN, or the Association of Southeast Asian Nations. After the prime minister's unsuccessful trip to India in 2018, it was also time to increase our engagement with that country. And finally, we needed to strengthen our existing close ties with Korea and Japan.

We also needed to build our expertise at Foreign Affairs. When it came to the Indo-Pacific, this was clearly not at the level it needed to be. While there were pockets of expertise, it was lacking in breadth and depth and would take many years to build. I put the blame for this squarely on the shoulders of politicians. It is our responsibility to establish priorities and provide direction. In the case of the Indo-Pacific, that direction had been missing for a couple of decades. After more than a century of focusing primarily on our most important trading partner and neighbour to the south, as well as our European partners across the Atlantic—also a source of trade and of so many of our early immigrants—it was now urgent for us to look hard across the Pacific. Globally speaking, the twentieth century had been dominated by the United States and Europe. The twenty-first century was becoming the Indo-Pacific century, and Canada's foreign policy had been slow to recognize it.

While it could have been made public about a year earlier, had there been the will to do so, Canada finally announced its Indo-Pacific strategy in November 2022, knowing that it had some work to do if we were to avoid losing out on trade and being marginalized on security-related matters. In my opinion, the major elements of the strategy were the same as those I had been working on with the department. It will now be important to see whether we follow through with all the steps the strategy entails.

In mid-March 2021, I received my first COVID-19 vaccine. It was the Covishield version, made in India at the Serum Institute, under licence from AstraZeneca. Because the delivery of other vaccines ordered by Canada would take time, given high global demand, our procurement minister, Anita Anand, had suggested we make a formal request to India to purchase vaccines from them. Working with Nadir Patel, our high commissioner in India, I spoke with External Affairs Minister Subrahmanyam Jaishankar, asking that Canada be allowed to purchase up to two million doses. Despite his repeated criticism to me that Canada was soft on Canadian Sikh extremists advocating for an independent state of Khalistan and that we needed to do more to protect Indian embassy and consular officials against protesting demonstrators, the minister agreed to my request. (Canada and other countries would subsequently provide some assistance to India when they were dealing with a massive outbreak of the COVID delta variant.)

As we moved into April, I began preparing for my first trip abroad, to attend a meeting of the G7 foreign affairs and international development ministers scheduled for early May in London, England. This meeting was intended to lay the groundwork for the G7 leaders meeting in Cornwall the following month. This would be my first opportunity to meet my counterparts face-to-face. My fellow international development minister, Karina Gould, would participate virtually.

Although the city looked a little deserted because of COVID measures, it was good to be back in London, a place I had called home for six years. Our meetings took place over three days, focusing on issues of common concern. Despite not being members of the G7, India, Australia, Korea, Brunei (chair of ASEAN), and South Africa had been invited to participate, and the European Union was present as an observer.

The fact that Joe Biden had replaced Donald Trump certainly contributed to a more constructive atmosphere, as was evident by Antony Blinken's engagement with the other G7 ministers. The interpersonal chemistry was excellent. In my opinion, the meetings were productive, resulting in widespread consensus on the wording of the communique that was subsequently issued (not always a straightforward task). Issues of particular concern on the agenda included: global security with a focus on the actions of China, Russia, and North Korea; climate change and sustainable development; poverty and food security; human rights, including the education of girls and empowering women; media freedom; and arbitrary detention in state-to-state relations.

India and South Africa would make strong pleas for G7 leadership in ensuring a more equitable distribution of COVID vaccines, given that poorer countries did not have the means to pay for them. They also argued that pharmaceutical companies like Pfizer and Moderna should make their intellectual property on mRNA vaccines public to allow others to produce them. This last proposal met with pushback for two reasons: the desire of pharmaceutical companies to protect their intellectual property, and the fact that setting up vaccine-making facilities was a much more complicated and lengthy process than many had thought, and it would not increase vaccine supply overnight. There was, however, consensus on the need for more equitable distribution, given the global nature of the problem. Canada would be one of the countries donating vaccines to other countries.

As is customary when ministers gather, they also engage in bilateral meetings with their counterparts. A priority on my agenda, given our decision to focus on the Indo-Pacific, was my meeting with Japanese foreign minister Toshimitsu Motegi to formally sign an agreement known as the "Shared Japan–Canada Priorities Contributing to a Free and Open Indo-Pacific," which focused on cooperating in the following areas of common concern: the Rule of Law; Peacekeeping Operations, Peacebuilding, and Humanitarian Assistance and Disaster Relief; Health Security and Responding to COVID-19; Energy Security; Free Trade Promotion and Trade Agreement Implementation; and finally, Environment and Climate Change.

I left London with a better understanding of the crucial importance of the personal interactions between ministers when discussing foreign policy. Canada can't go it alone. It must work with others if it is to accomplish its objectives, both domestic and international. That works best by having face-to-face meetings, including less formal moments of social interaction to share views, listen, and sometimes make compromises for the overall good.

My trip had sparked some media interest at home, not so much because of the issues discussed at the G7 but rather because it was the first time in many months that a Canadian minister was venturing abroad and there was intense interest in knowing whether I would follow all the COVID protocols before, during, and after my trip, which of course I did. I was tested before my departure and received a daily test while in London and again upon my return to Montreal, where I quarantined in an airport hotel until my results came back, followed by quarantine at home for a total of fourteen days. So, no story there!

Soon after, I was off again, this time to Reykjavik, Iceland, for a meeting of the Arctic Council. The eight members of the council are the countries that surround the Arctic Ocean: the United States, Russia, Canada, Finland, Norway, Sweden, Iceland, and Denmark (which includes Greenland). To quote from its website: "The Council's *Strategic Plan 2021–2030* guides its work towards the Arctic as a 'region of peace, stability and constructive cooperation, that is a vibrant, prosperous, sustainable and secure home for all its inhabitants, including Indigenous Peoples, and where their rights and wellbeing are respected." Living up to these obligations was a growing challenge in the face of climate change, resource exploitation, and increasing vessel traffic. In addition, although security and sovereignty were not part of the council's mandate, they were never far from anyone's mind.

During the visit, I held bilateral talks with each of my counterparts, among them Sergei Lavrov, the Russian minister of foreign affairs. Russia had just assumed the chairmanship of the council, and while one of my objectives was to reiterate the importance of our countries cooperating in the Arctic, I had other points to raise. Our discussion was frank. I had been warned that Lavrov could be a grumpy bear and I was prepared for that. He was also a man who could exert a certain charm if he chose to. I had seen this the night before, when all the ministers dined together. At our bilateral meeting, he spoke about how Canada and Russia had once been closer, and how it would be good for us to rekindle that friendship. He even mentioned the Murmansk Run, the hazardous maritime convoy route that saw Canadian merchant navy and Royal Canadian Navy ships sail into the Arctic Ocean to deliver supplies to Russia during the Second World War when the country was cut off by the German army. I was mildly impressed that he brought this up, no doubt because his office knew I had been in the navy. I took the opportunity to tell him that yes, indeed, we had once been closer—including in space, where we still cooperated—but that things had changed, primarily because of Russia's illegal annexation of Crimea and its continued destabilization of eastern Ukraine. (This was nine months before Putin's subsequent unprovoked and illegal invasion of the country.) Lavrov responded by presenting the standard Russian arguments to justify their invasion—bogus arguments that we have all, sadly, come to know so well.

During my visit, I was also able to meet with Antony Blinken again, where I took the opportunity to reiterate the importance of finding a way to get the two Michaels released from China. I raised this with him every time we spoke, and he clearly understood how important it was for Canada.

Returning home, I went into quarantine once again while preparations began for a twelve-day trip at the end of June, a trip that would take me to Italy for a meeting of the G20 foreign affairs and international development ministers, followed by a visit to Jordan, Israel, and the West Bank, and finally a stop in Lithuania for the Ukraine Reform Conference.

While I performed my duties, not a day went by that I didn't think about the two Michaels, unlawfully imprisoned and with no simple way to bring them back to Canada. Trying to secure their release required us to engage in back-channel diplomacy with a number of players, including the U.S. Department of State and the Department of Justice, senior Chinese officials in Beijing, and the president of Huawei, Ren Zhengfei, father of Meng Wanzhou; the objective was to find a solution that would be acceptable to all and enable the release of the two Michaels. Dominic Barton, our ambassador in Beijing, and Kirstin Hillman, our ambassador in Washington, worked non-stop on this, as did other officials in my department. At the request of Huawei,

I would also meet with senior representatives of the company in Ottawa prior to their own subsequent trip to Washington.

Periodically, I spoke with members of the two Michaels' families, to answer their questions and to reassure them that we were making every effort to secure the release of their loved ones, despite the appearance that no progress was being made. These phone calls were the most emotionally difficult part of my job. My objective was to communicate that we were making progress (which we were) and that they should not give up hope. This was challenging, since I could not discuss specifics. While some family members were understandably skeptical, or becoming so, they were invariably polite. I always came away with a profound sense of admiration for their courage in the face of the sheer hell they were going through.

Added to my conversations with the Kovrig and Spavor families were my equally difficult conversations with the families of the victims of Ukrainian International Airlines Flight 752. Their desire for justice was met by Iran's continuing refusal to be fully accountable for what they had done. The families were understandably exasperated and wanted Canada to do more. It was frustrating that all I could do was reaffirm our commitment to them and stand behind our promise not to quit until we achieved our goal.

On June 3, Canada, Ukraine, Sweden, and the U.K. delivered a notice of claim to the government of Iran concerning Flight 752. Our joint statement read: "We have made a series of demands that include, but is not limited to, an acknowledgement of wrongdoing and a full accounting of events that led to the downing, a public apology, the return of missing and stolen belongings of the victims, assurances of non-repetition in the most concrete terms, transparency in the criminal prosecutions in accordance with the rule of law, and equitable compensation for material and moral damages suffered by the victims and their families regardless of nationality and in an amount consistent with its obligations under international law."

Not surprisingly, as of this writing, Iran still refuses to negotiate. Other options to obtain justice are being pursued, in accordance with international law.

In late June, I flew to Rome to begin my twelve-day trip overseas. As I landed, a flood of memories resurfaced of my visit in 1964, when I was fifteen years old, staying in youth hostels and travelling by train through Italy with my brother Braun: how we saw Leonardo da Vinci's *Last Supper* in Milan, visited the Colosseum and St. Peter's Basilica in Rome, stayed in a medieval castle in La Spezia overlooking the Mediterranean and in a mansion in Florence that had belonged to Mussolini's mistress; how Braun was propositioned by a prostitute in Venice and I by a man in Rome; but most of all, how we looked after each other, as brothers do. Although Braun was gone, I still cherished those memories.

First on my agenda was a meeting of the eighty-six countries of the Coalition Against Daesh, the group more commonly known as ISIS or ISIL. (Daesh was an acronym of the group's Arabic name, which translates as "the Islamic State in Iraq and Syria.") It was a Salafi jihadist group, founded in 2004, which ruthlessly enforced its own strict interpretation of Islamic law. At its peak, it dominated large portions of eastern Syria and northwestern Iraq. Most will remember young Canadians being lured by its extreme ideology and recruited to join it.

Under the leadership of President Obama, the coalition was formed in 2014 and is committed to degrading and ultimately defeating Daesh. At our meeting on the outskirts of Rome, we all reaffirmed our resolve to ensure that the group could not resurface in the Middle East or secure a strong foothold elsewhere, as was beginning to happen. Unchecked, it could spread like a cancer in disaffected regions of the world such as Afghanistan and certain African countries.

To date, the coalition has made major progress in destroying ISIS in both Syria and Iraq. Beyond killing ISIS fighters, key leaders have also been eliminated, military infrastructure has been destroyed, and the command structure has been weakened, as has ISIS's ability to finance itself. That said, the risk of resurgence has not disappeared as thousands of fighters remain and ISIS continues to try to recruit new ones and reorganize itself.

In Africa, progress remains challenging as ISIS affiliates pop-up in regions such as the Sahel and West Africa, and in countries like Mozambique. Efforts to support those affected and to help them build their own counterterrorism capabilities are ongoing but will take time and a continued commitment.

From Rome, I flew to Bari, where I stayed overnight before the G20 meeting the next day in the historic city of Matera, a UNESCO World Heritage Site. Karina Gould, minister of international development, would join in via a link from Canada. Notably, the agenda focused on the required global response to the COVID-19 pandemic, including "equitable, worldwide access to diagnostics, therapeutics and vaccines." Italy, being the host country, also chose to shine the spotlight on sustainable development on the African continent. After a full day of covering a number of topics, the meeting ended with the adoption of the Matera Declaration, calling upon the international community "to build inclusive and resilient food chains and ensure adequate nutrition for all, in line with the 'Zero Hunger' goal set for 2030."

The meeting was a valuable experience that allowed me to appreciate the dynamics of this larger, more diversified group, compared to the G7. I was also able to hold a number of bilateral meetings with my counterparts. Whether the G20 resulted in concrete deliverables, and not just altruistic statements, is probably a matter of opinion depending on the issue, but the simple act of getting together and talking to each other made it worthwhile, especially during COVID and the various lockdowns. Dialogue keeps the channels open. That being said, when a group of countries includes China and Russia, achieving consensus can be challenging, to say the least. Fortunately, there is great value in interacting with such countries as Australia,

South Korea, Mexico, Turkey, India, Argentina, Indonesia, and others. Overall, the meeting went well, in part because the modest agenda stuck to subjects that enjoyed consensus.

From Italy I flew to the Middle East, beginning in Jordan, where I met Prime Minister Bisher Al-Khasawneh and Foreign Affairs minister Ayman Safadi. Given the constructive role that Jordan has long played in the region, including hosting a large number of Palestinian refugees, I wanted to discuss the recent unrest in Jerusalem, the violence at the al-Aqsa Mosque, the threatened expropriation of Palestinian homes from the Sheikh Jarrah neighbourhood of East Jerusalem, and the conflict between Israel and Hamas, which began with Hamas firing rockets into Israeli territory and Israel retaliating with air strikes on Gaza.

While in Jordan, I visited Jerash Camp (known locally as the Gaza Camp), the poorest of the country's ten Palestinian refugee camps. It dates back to 1968, when it was set up as an "emergency" camp for nearly twelve thousand Palestinian refugees and other persons displaced from Gaza by the 1967 Arab-Israeli war. Today, it is home to more than 35,000. I met with a group of six young students who spoke to me about their lives and aspirations. They had all been born in the camp. I also met a Palestinian family in their home, little more than a crowded shack where six children lived with their parents as the father struggled to find part-time work to feed his family. I gave them some local sweets as I departed and saw the children's eyes light up, a moment of happiness in otherwise hard lives.

Leaving Jordan, we drove to Jerusalem, where I met Prime Minister Naftali Bennett; Yair Lapid, my counterpart at Foreign Affairs; and Orit Farkash-Hacohen, minister of innovation, science, and technology. I also had a phone call with President Isaac Herzog. In addition, I had dinner with a group of thought leaders to hear their views on the current situation and possible future directions. Between these meetings, I spoke with representatives of various humanitarian organizations working in Gaza, dealing with the severe aftermath of the eleven-day conflict the previous May.

In my meeting with Minister Farkash-Hacohen, I confirmed Canada's interest in working with Israel in sectors such as space and artificial intelligence, a continuation of our past commitments to cooperate to our mutual benefit. In my one-on-one meeting with Minister Lapid, I reaffirmed Canada's enduring friendship and support for the State of Israel and our continued belief in the long-term goal of a two-state solution to achieve a lasting peace. I said that both sides in the latest conflict needed to take steps to "lower the temperature," given recent events and the fragile ceasefire. I encouraged Israel to cease both its planned evictions in East Jerusalem and its settlements in the West Bank, given their provocative nature. Lapid, while not responding to the specific points I raised, spoke about the need for calm on both sides and echoed my comments about the importance of Canada's friendship with Israel. I suspect he already knew Canada's position, so my words had not come as a surprise. Nevertheless, they needed to be said.

Forming the most recent government in Israel, with its many parties, had not been easy. Complicating matters was the fact that long-time prime minister Benjamin Netanyahu was splitting the vote between those who supported him and those who believed that he was guilty of corruption. After four elections, Naftali Bennett, leader of a small right-wing party, succeeded in forming a coalition and becoming prime minister, a job he would share with Lapid on an alternating basis, each assuming the position for two years.

My meeting with Prime Minister Bennett was quite formal, with officials from both countries sitting across from each other. On behalf of Canada, I congratulated him on forming the government and reaffirmed our enduring friendship, and in return, he thanked Canada for its long-standing support for the State of Israel. During our conversation, he mentioned that he had lived in Montreal for a short time as a young boy. I told him I represented a riding in Montreal with a Jewish population and two synagogues near my home. Normally, prime ministers meet with prime ministers and foreign ministers meet with foreign ministers. I considered myself fortunate to have met both, especially given that the new government was only a few weeks old.

My dinner with Jewish thought leaders was an opportunity for a wide-ranging discussion. It came as no surprise that I heard different views as to what the future holds for the region. What was also clear was my sense that support for a two-state solution had lost ground in the last decade. Whenever the subject was raised, the standard reply was that a two-state solution was all well and good but that it required someone across the negotiating table from Israel who legitimately spoke for all the Palestinian people, a non-starter with Hamas officially representing more than two million Palestinians in Gaza.

A remarkable moment for me while in Jerusalem was my visit to a school where Jewish and Palestinian children attended classes together, beginning at the elementary level. Instead of suspicion and hatred, what I witnessed was harmony. Like many, I'm sure, I could not help but wonder why these children and their teachers could not serve as an inspiration for others. The credit of course belonged to those who ran the school. The peaceful education of these children would stand in stark contrast to the horrific events that unfolded on October 7, 2023, and the many lives lost since then.

I was then driven to the West Bank, where I was met by Robin Wettlaufer, the head of the Representative Office of Canada to the Palestinian Authority, in Ramallah. The purpose of my visit was to meet with the president of the Palestinian Authority, Mahmoud Abbas, and his foreign minister, Riyad al-Maliki, as well as pay a courtesy visit to Prime Minister Mohammad Shtayyeh.

When I met Abbas, I reaffirmed Canada's commitment to a two-state solution. Abbas himself had participated in Israeli–Palestinian talks brokered by U.S. secretary of state John Kerry back in 2013 and 2014. At the time, there were real questions about how much power he actually wielded as the representative for the Palestinian side. The same thought went through my mind when I met him. That being said, he expressed frustration at Israel's relentless occupation of more and more land on the West Bank, an undeniable reality.

During our meeting, I took the opportunity to encourage him to hold long-overdue elections as soon as possible. This was important for the democratic process; many young Palestinians had never had the opportunity to vote, the previous elections having been held in 2006. As of this writing, those elections have yet to happen.

Finally, I expressed Canada's concern about a Palestinian activist, Nizar Banat, who had died recently in the custody of Palestinian security forces, and of the importance that justice be seen to be done expeditiously and transparently. While Abbas agreed with that, I was left uncertain whether he would do anything about it. Despite the ongoing trial of the fourteen officers implicated in Banat's arrest, as of this writing, justice has not been rendered.

While in the West Bank, I also visited the Palestine Techno Park at Birzeit University to observe the results of a Canadian investment to enhance economic opportunities for low-income women and youth. I also had dinner with Palestinian thought leaders, including Hanan Ashrawi, a well-known activist who has played important roles in the Palestinian Liberation Organization, the Palestinian Authority, and the Palestinian Legislative Council. She was joined by several young people who wanted to hear from Canada and share their personal aspirations for an independent Palestinian State, freed from the Israeli controls imposed on them in their daily lives. I reiterated Canada's unwavering support for a two-state solution, expressing the hope that negotiations could one day resume.

I left the Middle East with the sense that I had established personal contact with many of the key decision-makers in the region, a good first step. Unfortunately, I also left with the feeling that support for a two-state solution had decreased considerably and that something significant would have to happen to change that.

Since my visit, the political landscape has changed dramatically. Israel held another election and, once again, Benjamin Netanyahu became prime minister, having succeeded in forming a coalition with parties considered "far right" on Israel's political spectrum. More importantly, Hamas's brutal murder of 1,200 Israelis and capture of 240 hostages on October 7, 2023, triggered a chain reaction whose precise outcome is, as of this writing, uncertain. It has also reaffirmed for me the belief that only a two-state solution provides any hope of bringing lasting peace to the region, and that this will not happen as long as Palestinians allow Hamas to control Gaza.

What happens in the Middle East is felt beyond the Middle East. Canada experienced fallout from the turbulent events that occurred in Israel and Gaza just before my visit in 2021, as evidenced by spikes in anti-Semitism and Islamophobia. That fallout, however, was nothing compared with the demonstrations and violence in the aftermath of October 7. Aftershocks were felt all around the world, including in Canada. I have witnessed them more than once, within a block of my home in Montreal.

Before returning to Canada, I had one last stop to make, this one in Vilnius, Lithuania, to attend the Ukraine Reform Conference. This gathering focused on Ukraine's continued efforts to democratize its institutions and eliminate corruption. The meeting went well, and while there is still work to do, there was also consensus that reforms were moving in the right direction.

Canada has been contributing to these reforms since 1991 by supporting democratic development, including free and fair elections, and by helping strengthen civil society, media institutions, the rule of law, and judicial independence. Achieving this takes time in a country long controlled by oligarchs wielding disproportionate economic power and undue influence over the media, enabling them to control political outcomes. It requires an ongoing long-term effort to ensure the impartial application of the law, an unbiased media to hold governing authorities to account, and the creation of fair market competition. Canadian international development assistance programming has contributed significantly to Ukraine's recent progress in entrenching these crucial reforms, including measures to prevent corruption in key institutions.

During the visit, I met with several ministers, including Ukraine's foreign affairs minister, Dmytro Kuleba, with whom I had already had numerous conversations. It was an opportunity for me to reiterate our solidarity with the people of Ukraine. I also met Sviatlana Tsikhanouskaya, the leader of the Belarusian democratic movement who went into exile after dictator Alexander Lukashenko declared himself the winner of the 2020 presidential election. I had previously spoken to her by telephone, to assure her of Canada's unwavering support. Incidentally, during that conversation, she had asked me to pass on a message to Wayne Gretzky, requesting him to contact her. I surmise she was hoping that he might have a message of encouragement for the Belarusian people. She told me that he had ancestral roots in her country, where he was considered a hero. My staff left a message with Gretzky's office, although I don't know whether he followed up with a call to her. It was a pleasure to meet this courageous woman face-to-face. She was appreciative of Canada's support and of our imposition of sanctions on the despotic Lukashenko regime.

On a lighter note, I had a rather strange experience at the luncheon hosted by Lithuania. During the meal, waiters circulated behind the seated guests and offered them a choice of red or white wine. I say "them" since the waiter serving on my side of the table kept bypassing me for some reason. On his third pass, I turned around and asked him for some red wine. He looked nervous and whispered in my ear that he had been told not to serve me. Rather than making a fuss, I let it go. When we got to the dessert course, the same thing happened, with the waiters offering guests a choice of tea or coffee.

Again, the same waiter was bypassing me. On his third pass, I stopped him and asked for a cup of coffee. Looking nervous again, he whispered in my ear that he had been told that I could only be served a fruit tea. This time I asked to see the hotel's manager, who quickly arrived breathless. I stepped away from the table and asked her why I was being treated in this manner. She replied by asking me my name, which I gave her. She turned ashen and began to apologize profusely. Apparently, the seating layout had been changed at the last minute and I was seated at the place of a minister from another country who was not to be served wine or coffee. I couldn't help myself. I quickly scanned the table to see whether I could spot the minister in question, who no doubt would be in a very good mood, but he was no longer at the table. Later that afternoon, when I returned to my hotel room, I found an expensive bottle of champagne with a letter of apology from the manager. All in all, a funny story with a happy ending, but for a while I wondered what I had done that was so wrong that I couldn't have a glass of wine, let alone a coffee.

That evening, I joined my Canadian colleagues for a dinner, which included our ambassador to Ukraine, Larisa Galadza, and our ambassador to Latvia, Estonia, and Lithuania, Kevin Rex. It was an excellent opportunity for me to hear how Ukraine was doing under its new president, Volodymyr Zelensky, and to talk about Canada's military presence in Latvia. In both cases, the common threat was Vladimir Putin, in view of his annexation of Crimea and his support for separatists in the Donbas region of eastern Ukraine, and in the case of Latvia and other NATO countries bordering Russia, concern that he might do something similar. Less than a year later, of course, he would launch his assault on all of Ukraine.

It was clear from my discussion with Ambassador Galadza that Ukraine was making progress in its efforts to strengthen its governance and eliminate corruption, and also that President Zelenskyy was a popular leader. Ukraine also appreciated that Canada had been training Ukrainian soldiers since 2015, under Operation Unifier.

Similarly, Ambassador Rex confirmed to me that the people of Latvia were most grateful for the significant presence on their territory of a Canadian Forces battle group as part of Operation Reassurance.

The next morning, I flew back to Canada. It had, in my opinion, been a productive trip.

When Parliament broke for the summer, talk of an election was in the air, but no one knew for sure. If it happened, would it kick off in August or after Labour Day? Only the prime minister could make that decision. As I returned to my riding, I took a chance and rented a cottage in the Laurentians for ten days in early August. During that time, I maintained close contact with my department, given the rapidly evolving situation in Afghanistan.

August 15, 2021, turned out to be significant for two reasons: the writs of election were issued by the governor general, and almost halfway around the world, the Taliban marched into Kabul. In the case of the election, the prime minister was clearly hoping he would achieve a majority. As a precaution, I had taken steps to prepare, benefiting from a great campaign manager, Élisabeth d'Amours, who knew what she was doing. Thank God because, as it turned out, she and her team ran the campaign without me, while I focused 90 per cent of my time on a single issue: the debacle in Afghanistan.

Throughout the spring and summer, the probability that the Taliban would assume control in Afghanistan had increased significantly, given the well-publicized decision by the United States and NATO to pull out by September. Early intelligence suggested a Taliban takeover by the end of the year or perhaps late fall. As time went by, the predicted takeover moved closer to early fall (and ultimately late summer), due to growing doubts about the Afghan army's ability, or willingness, to defend their country. Consequently, the evacuation of Canadians, as well as Afghans who had worked for Canada, became increasingly urgent. I would spend most of my day working on nothing but the evacuation process, with daily briefings from my ministry and regular meetings with Ministers Harjit Sajjan and Marco Mendicino, of defence and immigration respectively.

Before I go into that, let me step back and talk about two under-utilized services that Global Affairs provides to Canadian citizens abroad: travel advisories and Registration of Canadians Abroad, or ROCA. The department publishes travel advisories on its website for Canadians abroad or those planning to go abroad. These are updated as required, either to advise Canadians not to travel to certain countries or to leave countries they are visiting because of increased concern for their safety. As the situation in Afghanistan deteriorated, the Taliban made it known that it might move into high gear after May 1 and attempt to take over the entire country. This could not have escaped the attention of Canadians in Afghanistan.

On April 30, Canada's travel advisory for Afghanistan stated: "If you're already in Afghanistan, you should leave." This was more than three months before the fall of Kabul. By July 15, the travel advisory stated: "If you're already in Afghanistan, you should leave immediately." This was still one month before the Taliban swept into Kabul. At the time, Kabul airport was still operating some commercial flights. Travel advisories on August 16, 17, and 18 reiterated the same message, with the addition of the words "if it's safe to do so."

The advisory issued on 26 August stated: "Canada's evacuation operations have ended. Until the security situation has stabilized, you should shelter in a safe place. Keep in mind that you are responsible for your own safety and that of your family."

For Canadians to benefit from their government's travel advisories, they must first see them. Many Canadians in Afghanistan either did not see them or chose to ignore them. With all the information you'd ever need at the push of a button on your cell phone, it's hard to fathom who would fall into the former category. This is comparable to Canadians holidaying on a Caribbean island and choosing to ignore hurricane warnings or taking the risk of staying, hoping that the hurricane will somehow avoid them. Clearly, we need to find a better way to ensure that Canadians travelling abroad monitor travel advisories, but Canadians also need to take some responsibility for their decisions if they choose to ignore warnings.

ROCA is another important service that is underused, with the result that it's impossible for the government to know exactly how many Canadians are in a particular country at any one time. Here is what the government website says: "Registration of Canadians Abroad is a free service that allows the Government of Canada to notify you in case of an emergency abroad or a personal emergency at home. The service also enables you to receive important information before or during a natural disaster or civil unrest. We encourage you to register whether you are planning a vacation or living abroad."

Once again, this service is of no value to Canadians who don't use it, and this became painfully obvious as the crisis in Afghanistan deepened and many Canadian citizens waited too long to evacuate. Because the advance of the Taliban occurred more rapidly than anyone expected, many Canadians visiting relatives contacted consular authorities only when it was too late and the Taliban had already taken over. It was only then that we discovered there were Canadians in Afghanistan whom we knew nothing about, as more and more of them made contact with Global Affairs for the first time, seeking help to get out. Again, we need to find a way to ensure that Canadians visiting countries with inherent risks attached to them take the simple step of registering their presence so they can be reached in the event of an emergency.

As I mentioned, based on intelligence reports in late spring, it was thought that a takeover of Afghanistan would not occur until well into the fall, or possibly not until the end of the year. Clearly that assessment was wrong, and it caught every single country, including Canada, off guard. Unfortunately, the Afghan army retreated hastily or capitulated, leaving the door open for the Taliban to enter Kabul in mid-August, at which point evacuation became extremely difficult.

Even they admitted being surprised at the speed with which they had taken over.

As the Taliban closed in on Kabul, the difficult decision was taken to close the Canadian Embassy. On the one hand, when you close an embassy, you end the services it provides. On the other hand, if you keep it open too long, you may be putting the lives of embassy staff at risk, which is inexcusable. For as long as diplomatic missions have existed, they have been potential targets, despite the immunities afforded them by the Vienna convention on diplomatic relations. We've seen it happen in many countries, and we've seen the horrible consequences. I announced the closing of the embassy on August 15.

Meanwhile, Canada had joined military airlift operations out of Kabul airport to evacuate as many Canadians and authorized Afghans as possible before the previously announced September deadline for U.S. and NATO troop withdrawal.

The road from Kabul to the airport was clogged with vehicles carrying people desperate to leave, including vehicles that had been deliberately abandoned by their occupants near Taliban checkpoints to avoid being arrested. It was pure chaos. These vehicles carried not only foreigners but also Afghans hell-bent on getting out of the country. If someone managed to make it to the airport, they still had to find the right people to let them through, as thousands tried to get inside the airport perimeter in the hope of getting on a flight. This led to chaotic and heartbreaking scenes of people standing in water trenches and begging to be let inside the guarded perimeter, and in some cases handing their children to soldiers in the hope that they at least might be evacuated. The whole world was horrified to see a desperate Afghan dangling from the undercarriage of an aircraft and falling to his death. Shortly after, a suicide bomber killed over 150 people just outside the airport, making matters infinitely worse. For over two weeks, we witnessed the desperation of people trying to escape Kabul, fearing retaliation from the Taliban or not wanting to remain in a country ruled by them.

During this time, Minister Sajjan was coordinating Canada's military airlift, Minister Mendicino was dealing with a growing number of Afghans seeking refuge in Canada, and I was working with other governments to obtain permission for Canadians and authorized Afghans to enter their countries as they fled.

Kuwait provided clearance for Canadian military aircraft to stage through their country and for passengers to be accommodated as they awaited onward travel. Pakistan agreed to allow Canadians and Afghans with the requisite paperwork to enter their country at specified land border points and to make their way to the Canadian High Commission in Islamabad for processing and onward travel. Getting to these border points was extremely risky, given the number of Taliban checkpoints along the way, but this option became one of the few available once the military airlift ended.

I was also in contact with Qatar's government to request their assistance in evacuating Canadians, which they generously agreed to help with. Because Qatar could speak to the Taliban, Qatar Airways had been allowed to schedule flights out of Kabul, and those accepted a number of evacuees, including Canadians. To cover all possibilities, I also asked my counterparts in Turkey and Tajikistan to temporarily accept any Canadians and Afghan nationals with the requisite paperwork trying to leave Afghanistan via their countries. To get to Turkey, evacuees would need to travel through Iran.

While efforts to evacuate Canadian citizens and permanent residents continued, parallel efforts were underway to evacuate Afghans who had been employed by Canada in various occupations, such as embassy staff or translators for the Canadian military, and who feared retribution from the Taliban. Appeals were coming in from many other groups, too, such as lawyers, judges, women activists, musicians, women soccer players, and others, to be allowed to come to Canada. I personally received many requests through concerned Canadian citizens wanting to help. Many Canadians, using their own savings, were helping fund the cost of Afghans staying in safe houses while they sought the paperwork required to leave the country. The government had committed to bringing them to Canada, recognizing the risk they ran if they remained.

Unfortunately, whereas the Taliban was generally willing to allow Canadians with passports to leave, they didn't feel the same way about Afghans, including those with the proper paperwork. Even Afghans with Canadian permanent resident status had no guarantee they could leave. The Taliban also made it difficult for children recently born in Afghanistan to a Canadian mother, but without proof of Canadian citizenship, to leave. Amidst this uncertainty, the Taliban kept moving the goalposts. They were clearly concerned with the prospect of a massive exodus of Afghan citizens.

Many of the Canadians helping to shelter Afghans in safe houses were former military who had served in Afghanistan and personally knew the people they were trying to help. They were clearly frustrated that Canada had not been able to find a way to get these people out. Had we acted earlier, they said, more could have been extracted before it was too late. In hindsight, it was hard to disagree. Unfortunately, we had miscalculated based on the intelligence we had at the time, and once the Taliban controlled the entire country, there was no magic solution. Not having an operating embassy also made the processing of applications extremely difficult, including the required background security checks and biometric data collection.

The government received requests to help pay for the safe house costs as donations from Canadians dried up. This was a reasonable request. Unfortunately, the government has to have proof that any money it authorizes will be spent as intended, requiring some sort of financial paper trail by those providing accommodation, a non-starter in a country in chaos and under siege. Should an exception have been made in this case? Some would argue yes.

Retired generals who had served in Afghanistan found a receptive audience in the Canadian public as well as media outlets willing to allow them to vent their frustrations nightly and complain that the

Canadian government was not doing anything to help. Their voices were amplified by those of Canadians involved in separate efforts to bring Afghans to Canada.

In hindsight, I would say this: the announced departure of U.S. and other troops by September sent the Taliban into high gear, resulting in the complete takeover of the country more quickly than anyone anticipated given that the allied forces' intelligence had assumed a slower advance and greater resistance by the Afghan army, an assumption that proved false. This collective failure was the single most important reason for the ensuing chaos and the failure to evacuate Canadians and Afghans seeking refugee status before the country was overrun.

In the case of Canadian citizens still in Afghanistan when the Taliban took over, failure to consult or to heed the travel advisories in the months preceding the takeover, and failure to register through the ROCA service, also accounted for some Canadians being stranded. Once commercial flights out of Kabul ceased, there were only two options left: the military airlift, which lasted from 4 to 26 August, evacuating 3,700 people, or escaping across a land border, most often through Pakistan. In the case of Afghans with Canadian ties, a perfect storm impeded their escape: the closing of the Canadian Embassy for safety reasons, their being forced into hiding out of fear for their lives, and, finally, the risks involved in crossing a land border, with Taliban checkpoints everywhere.

Under normal circumstances, Canada requires background security checks, including biometric data, before allowing a foreign national to come to Canada to become a citizen. Most would argue that this is a necessary measure. Could the rules have been relaxed in the case of Afghans, given their urgent circumstances? There are those who thought that this should have been the case. Had that happened, though, an important obstacle would still have remained for those Afghans: getting out of the country.

———

On September 20, 2021, Canadians elected another Liberal government, with almost the same seat count as before. The prime minister had gambled on winning a majority and lost. Although I dutifully defended his decision to call it, I was not happy that he had done so, given that we were dealing with a major debacle in Afghanistan and that we were not yet out of the woods with COVID. Less than two years since the previous election, there was no pressing need to go to the polls, interrupting the business of the nation for several months, not to mention at great cost to the taxpayer.

Before the election, our government bent over backwards to please the Quebec nationalist government of François Legault. This was a deliberate strategy with the hope of gaining enough Quebec seats to secure a majority. While all parties are guilty of blatant political pandering, it is not how a federal government should conduct itself in the interests of all Canadians. I had told some of my Quebec colleagues that it was naive to think that placating Legault on a number of files would stop him from publicly throwing his support behind the Conservatives, which is what he ended up doing, announcing that a Conservative victory would be better for the interests of Quebec.

In the weeks that followed, I continued to focus on Afghanistan. This included planning a trip to Pakistan, Kuwait, and Qatar, important countries in the region with whom we had worked. While Canada had initially committed to bringing twenty thousand Afghans to Canada, it decided to raise that number to forty thousand. Needless to say, a great deal of work remained to be done.

Because I had been in the job for only seven months when the election was called, I had made the difficult decision to run again, despite my family hoping I would retire and my promise to them after the October 2019 election to do so. I hadn't known at the time that I would be moved from Transport, and I felt that Foreign Affairs needed long-term stability and continuity, with still so much to do. In addition, the prime minister had not given me any indication that I would not be reappointed to the post following the election. That happened in early

October when he called to inform me that I would no longer be in cabinet. He did not offer an explanation. Had he told me this before the election, I would not have run.

It felt like a punch in the gut. There's no other way to put it: I had been dismissed. Saying I was disappointed does not convey the intensity of my emotions at that moment. I felt totally blindsided and didn't know what to think. Had I done something wrong, and if so, what? Was I taking the fall for Canada's response in Afghanistan or was it simply poor chemistry between me and the prime minister? Only he can answer that question, and he chose not to do so. The PM does not owe me an explanation, and I felt it was inappropriate for me to ask him why. I will say that none of the speculative reasons offered by friends and colleagues were related to my competence. Nor was I ever disloyal to him. Of course, that did not stop a few in the media from speculating about my removal. There's nothing you can do about that. It comes with the territory.

But whatever the reason, I was a big boy and I had to accept it and move on, which I would do in the days that followed. Prime ministers choose their cabinet and can dismiss their ministers without explanation whenever they choose. When Trudeau called me to announce his decision, he offered me the job of ambassador to France. It felt like being thrown a bone. For several reasons, I declined. I will elaborate on this in the next chapter.

As I've noted before, it was abundantly clear that I was not in the prime minister's inner circle and that our interpersonal relationship was minimal at best. However, I was a minister with two major files that needed attention from the top, and for whatever reasons, it was just not there. Sometimes I did not agree with his decisions, and that's fine. It happens. Our personalities were quite different, as were our tastes (after all, there was a twenty-three-year age difference between us), and there was nothing either of us could do about that. When all is said and done, I didn't have to be his close friend to do my job competently, although I do admit to wondering why he had assigned me two important portfolios and then devoted so little time to establishing a more personal working relationship with me.

Life has special moments even in difficult times. Although I was minister of foreign affairs for only nine months, I would participate in two memorable events before my departure. The first was the release of Michael Kovrig and Michael Spavor.

On September 24, I was with the prime minister when he announced that the two Michaels had been released and were on their way home. I had been briefed on this possibility in the days leading up to the announcement and the all-important choreography to ensure coordinated releases on both sides. The release of the Michaels was timed with that of Meng Wanzhou, who was now free to leave Canada after the United States dropped the request to extradite as part of her deferred prosecution agreement with the U.S. Department of Justice. Both Michaels arrived at Calgary airport early the next morning, where the PM and I greeted them on the tarmac. Their ordeal had lasted 1,019 days.

I had the opportunity to speak to both of them and to listen as they spoke to others. Conversations often start out with the niceties: How was the flight and how are you feeling? They looked surprisingly fresh considering the long distance they had travelled and the flurry of events leading up to their release. While I mostly wanted to hear what they had to say, I did mention my calls with their families as well as some face-to-face meetings. This was not a moment to start asking questions about their treatment by the Chinese authorities and their prison conditions. This was a moment for all of Canada to rejoice.

Although their personalities are quite different, the two men both exuded an inner strength and calm that was remarkable given what they had endured. They were two incredibly resilient human beings. They had been severely tested and had found a way to deal with it, unvanquished.

Many will write about how and why they were finally released, and each account will fail to provide the whole story. It is in our nature to seek simple explanations. The truth is much more complicated and involves a much larger cast of actors, each playing a role, each contributing—contributions that may seem inconsequential at the time but that in the end move us one inch at a time towards the final outcome.

My last proud moment as minister of foreign affairs came a few days later, when I spoke at the United Nations General Assembly in New York City. I viewed it as a great honour to speak on behalf of Canada. It was also a pleasure to be hosted by Canada's ambassador to the UN, Bob Rae, and his wife, Arlene. Through our years as colleagues, I had always valued Bob and Arlene's friendship and sage advice. They also hosted a dinner for me in their home so that I could meet several UN ambassadors to discuss Afghanistan.

Truth be told, I have always had mixed feelings about the UN, particularly the veto rights of the five permanent members of the Security Council (China, Russia, the United Kingdom, France, and the United States). I also find it abhorrent that certain repressive countries are given responsibilities related to human rights. Still, I remain convinced that it's important for Canada to support this world body, mainly because there are no other alternatives of this scope and size. Because we all share planet Earth and the global problems that confront us, we're in this together and have no choice but to act in unison to ensure the future of humanity. As I write this, the Israeli–Gaza conflict rages on, the war in Ukraine has passed the two-year mark and shows little sign of slowing down, the climate is changing at an alarming rate, we are recovering from a global pandemic, mass refugee migrations are happening in Africa and the Americas, and unending atrocities are being committed in countries such as Ethiopia, Sudan, and Myanmar. Obviously, it would be helpful if the UN had more power and ability to influence such events and wasn't as bogged down by internecine politics and the actions of certain members to undermine the good work that could be achieved. But it's all we have and, at least in principle, it's what is needed to help bring us together to fight existential threats now and in the future.

Still, knowing I would never have another opportunity to address such an international audience, I decided, with Bob Rae's encouragement, to personalize the opening part of my speech. Allow me to indulge myself and present it here.

As I address you, I am conscious that I am speaking to virtually the entire world. In my previous career, I was an astronaut, and I had the opportunity to see the entire world from the vantage point of space.

I have flown over all your countries, and I have reflected a great deal on our planet, Earth.

I have realized that Earth is the cradle of all humanity, and that we all come from the same place and that we have nowhere else to go, and that we must find a way to get along with each other, and that we must take care of our planet—a planet that we are visibly damaging.

Space offers the unique perspective of seeing beyond one's own national borders.

In that sense, this body, the United Nations, offers that same perspective.

I am honoured to be here with you today on behalf of Canada's newly re-elected government, led by Prime Minister Justin Trudeau.

I would like to begin my address by respectfully acknowledging that the land on which we gather today is the traditional unceded territory of the Lenape people.

Fellow delegates, friends—we are assembled today at one of the most challenging times in generations.

The world is facing simultaneous and cascading crises, including climate change, the COVID-19 pandemic, and threats to international peace and security, which serve to exacerbate inequalities, test our resilience, and shine a bright light on the shortcomings of our systems and institutions.

But not for the first time has this institution faced such formidable challenges. We must not be querulous or faint-hearted in the face of the hardships and difficulties of our modern world. That is not why we are here.

We must learn from the vision and courage of those who have gone before, and we must think of the hopes and aspirations of those who will inherit the world we leave behind.

From the ashes of World War Two, our parents and grandparents responded to the unprecedented social and economic collapse of the 1930s and '40s, with its accompanying death and destruction, by building a new international order based on rules and strong international institutions to bring stability, prosperity, and peace for the generations that would follow. They did not wring their hands in despair. They rolled up their sleeves and went to work.

Climate change, COVID-19, the rise of authoritarianism and inequality—these are the challenges of our time. They are ours to solve and overcome.

In doing so, we must look toward the future with optimism. Just as our parents and grandparents stepped up to the challenges of their moment, so too must we recognize and seize our own opportunity to shape the future . . .

EIGHTEEN

IN A SENSE, my political career had come full circle. After failing on my first attempt, I was elected in 2008 and proudly took my seat in the last row of the opposition benches. For the next seven years, my party would toil in the wilderness, first as the official Opposition, then as the third party. During that time, I focused on learning my job and gradually assumed more responsibility. In 2015 we broke through, and I became a cabinet minister, assuming two portfolios over six years. Now I was returning to the backbenches, although I remained on the government side of the House. Apart from possible committee work, I was now unencumbered by other responsibilities unless I volunteered to take them on, and I could focus more on being the MP for NDG-Westmount.

I was often asked why I turned down the PM's offer of the ambassadorship to France. I can assure you it wasn't that I didn't like Paris. Everyone loves Paris, and Canada's relations with France are certainly important. One reason was that my wife's parents were at a stage in their lives when Pam wanted to visit them frequently and not be too far away in the event of an emergency. The other was that I felt that the offer was not a strong enough reason for me to step down from my job as an MP, having been re-elected only weeks before. If I was to leave political life so soon after being re-elected, it had to be for a reason I could justify to myself, and in my view there was only one such reason (other than my health or a decision to retire for personal reasons), and that would have been to become the Canadian ambassador to the United States, a job for which I felt qualified, based on my previous career and responsibilities. I raised it with the prime minister, for his consideration when the current ambassador had finished her term. He had not expected that from me. He said he would get back to me. A week later he called to tell me it was not an option he was prepared to consider.

For the record, I was offered another ambassadorship in late 2022, as well as a third in 2023. I said no to both.

As Canada's forty-fourth Parliament took shape, I was given a slot at the Montreal General Hospital to get my left hip replaced. Such slots were rare, given COVID-19, and I jumped at it. My hip had been nagging me for some time and preventing me from walking or cycling without discomfort. I had abused my body with hard sports and that was now catching up with me. I don't remember a thing about the operation, even though I was not given a full anaesthetic. My wife Pam, who had scrubbed in on such operations as a nurse, told me it was not uncommon for patients to report hearing "carpentry" noises during the procedure, but gladly I didn't hear a thing. Carpentry noises are definitely not something I would want to hear while I was being operated on. A couple of hours after I woke up in the recovery room, I went for a test drive with my walker through the corridors of the hospital, and they also made me walk up and down some stairs with the help of a cane. Having passed the test, I was cleared to go home the same day and soon after started my rehabilitation. A few weeks later, a medical technician removed the nineteen surgical staples used to close the opening in my hip.

If I can offer any advice to those of you who might one day undergo the same surgery, it is this: do your physio religiously before the operation and as soon as you can afterwards, even though it's hard at the beginning. It will make a big difference. Within a few weeks I was back to driving my car. I felt like a new man. With a new lens in my right eye following cataract surgery earlier in the year, cow bone implants in my neck thirty years earlier, and now a left hip replacement, I appear to be batting 100 per cent on my body parts replacement program. Steve Austin, the Six Million Dollar Man, had nothing on me.

As we recessed for the Christmas break, I wondered how exactly I could contribute to Parliament. As is customary, our party whip asked each Liberal MP who was not in cabinet or a parliamentary secretary which standing committee he or she wanted to be on. After some thought, I decided to ask for something challenging and new to me: the Indigenous and Northern Affairs Committee. I saw it as a way of contributing to the ongoing process of reconciliation. Beyond the legacy of residential schools, there were so many other problems to fix, beginning with the provision of adequate housing, schooling, and health care. The list was long. It was my hope that this committee could, by undertaking focused studies on specific issues, make useful recommendations to the government.

My choice was approved by the whip, who asked me to chair the committee, which would be a new experience for me. It's a job that requires a thorough knowledge of procedures and, equally important, tact, diplomacy, and a certain level of neutrality.

Being a backbencher (even though I sat in the front row) allowed me to think more broadly about my life, and most importantly, about the future. As a cabinet minister, I had rarely had a chance to look beyond the coming week. For the first time, I did not feel there was another mountain for me to climb. I was now more interested in how to spend time with my family and friends, while continuing to work on the things that gave me personal satisfaction. I was ready to climb some smaller hills, but only the ones that interested me.

In my somewhat new role as MP Garneau, I was asked by various colleagues to take on new tasks, on the assumption that I had more free time. The requests were usually couched in the following terms: "Given your experience with this particular issue, could we ask you to do such and such?" Notwithstanding the flattering pitches, I decided that I would only agree to tasks that really motivated me and where I felt I could actually make a difference.

Early in 2022, Senator Yuen Pau Woo approached me with an interesting proposal: Would I be interested in joining an informal parliamentary group of two senators and three MPs to travel to Seoul, South Korea, for a week of meetings with Korean legislators and ex-politicians (including three former prime ministers and the former UN secretary general Ban Ki-moon), as well as scholars, think-tank leaders, and other stakeholders? We would be in the company of Dr. Kyung-Ae Park, who was organizing the trip. Dr. Park had initiated the Knowledge Partnership Program some eleven years prior at the University of British Columbia, with the purpose of engaging North Korean officials in constructive track-two, or back-channel, diplomacy. She achieved this by hosting chosen officials at UBC, generally for six-month periods, which allowed them to see how Canada managed economic sectors such as agriculture, forestry, and tourism. This form of "knowledge diplomacy" was aimed at keeping the door slightly ajar with a country that was otherwise almost totally isolated from the rest of the world. Up to then, about fifty North Korean officials had visited Canada.

The Republic of Korea is a remarkable country. Having been annexed by Japan in 1910 and subjected to brutal colonialism until 1945, it then had to wage a devastating war with North Korea from 1950 until 1953. Rising from the ashes of that conflict, it has gone on to become an economic powerhouse, the thirteenth largest economy in the world based on 2022 World Bank GDP rankings, chiefly because of a well-coordinated effort between the private sector and the government to build a strong manufacturing base, allowing it to export its products and create wealth. Huge conglomerates such as Samsung and LG are known the world over, as is "K-culture," which encompasses music, the film industry, fashion, and food. Canada

and Korea share a great deal in common. With similar-sized econo-
mies, strong democratic principles, and respect for the international
rules-based order, we are natural allies.

Because of my strong interest in the Indo-Pacific region, I accepted
Senator Woo's invitation without hesitation. The week I spent in Seoul
was, without a doubt, one of the most stimulating and enjoyable
weeks of my political career. We had access to some influential
Koreans and had numerous constructive exchanges with them. That
such an important group of individuals was willing to meet us was a
testament to the high regard in which they held Dr. Park and her
Knowledge Partnership Program.

The main focus of our visit was the issue of peace and security on
the Korean peninsula, given North Korea's nuclear weapons and
ICBM programs and the fact that the reunification of the two Koreas
is not on the foreseeable horizon. It's important for Canadians to
fully appreciate the politics of the region. Unlike Canada, which has
the luxury of living next to a friendly neighbour and ally, which also
happens to be the most powerful country on Earth, Korea lives with
the existential threat posed by its rogue and unpredictable northern
neighbour, not to mention its location close to both China and
Russia—a completely different reality from ours.

In April 2022, I was asked to co-chair the special joint parliamentary committee reviewing medical assistance in dying, or MAID.
The committee had been created the year before with a slightly
different membership. I agreed to this, given the importance of
the issue to potentially every Canadian citizen. The committee was
comprised of five senators and ten members of Parliament (five
Liberals, three Conservatives, one NDP, and one Bloc Québécois).
My co-chair was Conservative senator Yonah Martin.

Legislation modifying the Criminal Code to enable MAID had
already been passed, and our task in reviewing it was to focus on five
issues: MAID requests in the case of patients with a mental illness

as the sole underlying medical condition; whether MAID requests should be allowed in the case of mature minors (children who are deemed sufficiently mature to make their own treatment decisions); whether MAID advance requests should be allowed and, if so, under what circumstances; the state of palliative care in Canada; and the protection, with respect to MAID, of Canadians with disabilities.

There is no question that MAID is an emotional issue. It absolutely should be, considering what is at stake. Some people are resolutely against it, even when all medical treatments have been exhausted and a person is living in constant, unbearable, and irremediable pain. Others believe it allows a person to end their life with dignity, at a chosen time, surrounded by their loved ones.

We would hear from witnesses for seven months and receive many written submissions. I knew that our committee's final report had to be the fruit of careful consideration because of its bearing on so many people. As a co-chair, it was important for me to allow all sides to be heard. That said, I had my own views, as would any human contemplating the question. Witnessing my own father's last two painful years of life suffering from Lewy body dementia had made me a supporter of advance requests. I also endorse the view that a person under eighteen, suffering from unrelenting and incurable physical pain and facing foreseeable death, as well as having the full capacity to understand the implications of requesting MAID, should be allowed to receive it.

I am also supportive of MAID for a person suffering from a mental disorder as the sole underlying medical condition, provided that the person has been suffering relentlessly, without relief, for a long time and that every available treatment option has been tried without success.

On the question of MAID in the case of people living with disabilities, while I support it for people who request it and satisfy the legal requirements to access it, I also recognize the need for society to focus greater attention on structural problems such as poverty, isolation, and loneliness, which disproportionately affect persons with disabilities and may be contributing factors in their request to receive MAID.

My views about MAID are based on two beliefs: first and foremost, that society must respect the rights of individuals who satisfy the conditions required to receive MAID and request it; and that individuals should be given the choice to die in dignity.

On April 5, 2022, my good friend Bjarni Tryggvason left us. He was seventy-six and his death came as a shock. He had been one of the original six Canadian astronauts chosen in 1983. He flew on shuttle mission STS-85 in August 1997. I had seen him a few weeks before his death, when the Canadian astronauts met for dinner. He was still flying various types of aircraft and had a long list of projects he still wanted to do. There had been no hint of a health problem.

He was a dear friend for thirty-eight years and the smartest engineer I ever met. He understood everything there was to know about how airplanes fly. He was also an accomplished pilot, who spent many hours teaching me the basics of flight. He loved flying aerobatic aircraft and occasionally took me into the sky to do acrobatics until I was so disoriented, I begged him to stop. He was proud of his Icelandic roots, and when I was in Iceland for the Arctic Council, everybody there knew about him. When he flew in space, he tested a device he and his team had designed: a vibration-isolation mount that allowed experiments to be performed in near-perfect weightlessness, without the low-level vibration that can be transmitted (via motors, fans, and other moving parts) through the equipment rack housing the experiments. It is still being used. Bjarni was truly an original, someone who will never be replicated. He had lived his entire life in top gear. I was able to pay tribute to him in the House of Commons a month after his death, and all members stood up out of respect. Later in the year, I and many of his long-time colleagues gathered at the Canadian Space Agency with his family to plant a tree in his name.

When people who have marked your life die, as happened in my case with Jacqueline, Braun, my father, my mother, and now Bjarni, you can't help but reflect. Why is it that things have to change? For thirty-eight years, we, the original six Canadian astronauts, had been there, a constant in the lives of Canadians, with no apparent end in sight. Now, sadly, one of us was no longer there, a jarring reminder that nothing is forever and an admonition to savour every moment.

While my committee work was immensely satisfying, I was now giving serious consideration to my next decision. Since beginning my professional journey at sixteen, when I entered military college as a naval recruit, I had not stopped—for fifty-eight years. I was content with what I had accomplished, but now wanted to spend more time in simple contemplation.

That fall, I made the decision to step down from politics after tabling the MAID committee report in February 2023. I had discussed this with my family, and everyone was on board. You might even say they looked relieved! The moment was right. I hoped my constituents would forgive me for not serving until the next election. In the meantime, there was still work to do and I did not intend to relax the pace.

That fall, an unexpected issue seized me: the government's legislation to modernize the Official Languages Act. In my opinion, this legislation, known as bill C-13, was flawed because it included references to Quebec's Charter of the French Language in what was a federal bill. For me, this was unacceptable because it gave the charter de facto legitimacy, which had implications if it were to be challenged as discriminatory with respect to the language rights of Quebec's anglophone minority. I decided to appear before the parliamentary Standing Committee on Official Languages to voice my concerns. Although I spoke in French, this is the English translation that appears in the official record of the committee's proceedings of December 13, 2022:

I will begin by saying that Bill C-13 deals with a federal act that concerns the official languages of Canada, obviously. In my opinion, it is not appropriate to refer to Quebec's Charter of the French Language in Bill C-13, which falls under federal jurisdiction and deals with official languages in Canada.

By making this reference, we are de facto incorporating the Charter of the French Language of Quebec in a federal statute.

Let me remind you that Quebec's Charter of the French Language is not just simply Bill 101, which we have lived with for a very long time. It is now an amended charter by virtue of Bill 96. Yes, Bill 96 seeks to protect French in Quebec, which is a good thing, but it also discriminates against the anglophone minority.

What's more, Bill 96 also invokes the notwithstanding clause as a preventative measure, which creates many problems. It's as if we are saying that we will not entertain any argument or claim that calls into question, for whatever reason, the Charter of the French Language or Bill 96.

I hope that we all recognize, as federal MPs sitting on a federal committee and considering a federal act, that it would be a huge error to give Quebec free rein to do what it wants in linguistic matters in Quebec.

As federal MPs, we have a duty towards linguistic minorities in Canada, including Quebec's anglophones.

I was disappointed that my own party was responsible for enshrining references to the Quebec charter in federal bill C-13 even though my colleague Anthony Housefather and I asked the prime minister and minister Ginette Petitpas Taylor to remove them, something they were unwilling to do. In my opinion, this created the risk that, should the charter be challenged in court, the province could simply point to the fact that it was recognized in Canada's Official Languages Act. As a legislator, this became my hill to die on, and I said so publicly.

It is my firm belief that French needs to be protected in Canada, including in Quebec, but not by discriminating against the anglophone minority in the province. I had been a strong supporter of Bill 101, Quebec's original charter introduced in 1977, but I felt that Bill 96, introduced in 2022 to update the charter, went too far in its concessions to francophone Quebec, not to mention its invoking of the notwithstanding clause. I was concerned with the erosion of existing government services in English, the potential cost burden to anglophones in translating of documents, the added workload that could be imposed on English businesses, and the impact on student enrolment in English CEGEPs. (My riding was home to two English CEGEPs, Dawson College being the largest in the province.) Premier Legault loved to say that English is not in peril, and I agree, but that was not the point. It was the rights of anglophones in Quebec that were at risk.

Bill 96 had caused serious concerns in Montreal's anglophone community, including my own riding, and had led to several protest marches and a great deal of media coverage, primarily in Quebec. Since then, the Quebec government has begun to provide additional "clarifications" on how its charter will be applied. There has even been talk of "exceptions." Meanwhile legal challenges are underway.

In January 2023, the Special Joint Committee on MAID received the draft report that had been prepared by our analysts, based on the testimony of witnesses and the many written briefs submitted to the committee. In the following weeks, we finalized the report, including its recommendations. I had the honour of tabling it in the House of Commons on February 15 while my co-chair, Senator Martin, did the same in the Senate.

Co-chairing the MAID committee was one of the most demanding tasks I undertook in my fourteen years as an MP, both intellectually and emotionally. I was acutely aware of the fact that we were delivering groundbreaking recommendations to the government, most not based on unanimous agreement, but rather on the opinion of the majority. Chief among them was support for advance requests, a measure favoured by more than 80 per cent of Canadians, particularly in cases where a person received an initial diagnosis of dementia and would eventually lose the capacity to request MAID due to their illness. The committee also supported allowing mature minors to request MAID in Track 1 cases, where death was considered reasonably foreseeable and the minor was judged to have a capacity equivalent to that of an adult to make an informed decision. We also recognized the need to address structural factors such as poverty and isolation so that they did not contribute to a MAID-qualified person's decision to ask for it. Furthermore, we recommended that more palliative care be made available to Canadians, a provincial responsibility. The Conservatives, meanwhile, issued a dissenting report.

In submitting its report, the committee also expressed its support for the government's decision to delay by one year (until March 17, 2024) access to MAID in the case of a mental disorder as the sole underlying medical condition, in order to ensure that all the necessary safeguards, training, and oversight were fully developed and in place.

Several months after we issued our report, the government accepted some of our recommendations, but also decided that Canada was not yet ready to go ahead with advance requests or allowing mature minors to request MAID. (In January 2024, it also announced that it would not move forward on access to MAID in the case of a mental disorder.) In my opinion, the decision not to adopt certain recommendations (for the time being) was driven in part by political considerations. It was simpler for the government to say that there was more work to do, thereby avoiding the fallout from a decision to proceed (which it considered greater than the fallout from not proceeding).

It's instructive to remember that in the early 1990s, when Sue Rodriguez first challenged the courts to allow MAID, invoking section 7 of the Charter of Rights and Freedoms, her case went all the way to the Supreme Court and was defeated five to four. Twenty-two years later, in 2015, the Supreme Court ruled nine to zero in the Carter case to allow MAID in the case of a competent adult suffering a grievous and irremediable illness and where natural death was reasonably foreseeable. This led to federal bill C-14. In 2019, the Truchon-Gladu decision in Quebec stated, on the grounds that it was unconstitutional, that MAID could not be withheld in the case where a person's death was not judged to be reasonably foreseeable, which led the federal government to adopt further legislation (bill C-7). The point is that MAID came about not because governments wanted it, but because of court challenges. Society changes, and while governments and the courts should proceed cautiously on matters like MAID, more changes are probable over time.

On March 6, two days before I retired from political life, Ken Money, another one of the original six Canadian astronauts, passed away at the age of eighty-eight. His health had been declining for several years. Ken was the one who never got to fly. His time had simply run out. I know he would have done a great job had he been given the opportunity. When he was chosen, he was already an internationally recognized expert in vestibular physiology, a pilot in the RCAF, and an outstanding athlete who'd placed fifth in the high jump at the 1956 Olympic Games in Melbourne, Australia. He was an exceptional human being whose life had intersected with mine and enriched it.

The time had come for me to keep my promise to my family. I wanted to retire the right way and, fortunately, my former chief of staff, Marc Roy, offered to help. First, I would say goodbye to my staff in Montreal and Ottawa, people who had served me faithfully throughout my career. I would then notify the prime minister and the Speaker of the House. Next, I would bid farewell to the Quebec caucus, followed by the national caucus, and, finally, I would say a few words in the House, with my family and a few friends in the gallery. I was also willing to answer questions from the media on my way out the door, the main one being: Why are you retiring? I suppose it was a compliment to be asked that, rather than the opposite: What took you so long?

After all that, I would clear out both of my offices, repositories of a great deal of material accumulated over the years, which Pamela made a point of telling me not to ship over to our already crowded house. I would also begin to disengage from the government, which was no longer my employer.

It all unfolded as planned and I retired officially at the close of business on March 8, 2023. My NDG-Westmount riding association held a farewell event for me a few days later, which gave me the opportunity to thank the many friends and volunteers who had supported me for so many years. It was an emotional time.

As I reflect on my fourteen years as an MP, I think it's fair to say the job of a politician is rewarding but also demanding, even frustrating at times. The hours are long, and it's hard being away from family for half the year, particularly if you're needed at home for an urgent personal matter. Living out of a suitcase also becomes tiresome. The job can change unexpectedly at any time, and there is the need to get re-elected every few years. Mounting an election campaign and then running it is also hard work. Finally, putting up with personal attacks and so much negativity, especially in social media, can wear you down if you allow it to get to you.

The positives, though, far outweigh the negatives. I was given the incredible privilege of serving my country: debating and voting in Canada's House of Commons, running two departments with great teams and getting important programs underway, travelling to interesting countries, and visiting other parts of Canada, including our spectacular North. All these wonderful experiences more than made up for the demands of the job. Even my last stint as a backbencher was more than rewarding. I wouldn't have missed it for the world.

My time in politics taught me that it was possible to disagree and to do so respectfully. I witnessed this during my time at Foreign Affairs in my dealings with other countries. Remaining respectful can defuse a difficult situation and keep the door open to finding a solution. The same applies to discourse in the House of Commons.

As parliamentarians, we live in glass houses, always in full view of the public. You would think this would curb displays of bad behaviour. Surprisingly, that is not always the case, despite the bad example it sets and the public cynicism it creates. And yet, I and my colleagues were perfectly capable of behaving well when we chose to.

In my last speech in the House of Commons, I left my colleagues with the following parting words.

Before I finish, let me issue a challenge to everyone in this chamber. Arrive each day in this House with the firm intention of showing respect for colleagues and for this extraordinary place. Be dignified. We must remind ourselves that when emotions run high, as they do for all of us, those emotions need to be channelled in a positive way, whether when supporting something or criticizing it.

We all know that we are capable of dignified behaviour. We all know that we are capable of being critical without resorting to yelling at the top of our lungs. We all know that we want to be heard and even listened to when we ask a question or give an answer. God knows that the Speaker of the House reminds us of this often.

My challenge to members is to find their better angels and put away the anger and false indignation. Criticize by all means, but do it with respect and maybe even wit. Make Canadians proud of this House and the people in it. . . .

Now it is time for me to go. It has been an honour serving my country alongside all members.

Did I do everything I wanted to do? In politics, as in other occupations, you set your sights on getting certain things done, knowing that, on the day you leave, you will not have succeeded in every case. I spoke earlier of my hopes to do more at Foreign Affairs and I mentioned my dream of helping to create a Children's commissioner, given that we had ratified the UN Convention on the Rights of the Child. I had twice introduced a private member's bill without success when in opposition. I had resolved to pursue this matter if we became the government. When that happened in 2015, I tried to make my case to my party, knowing that it had previously voted in favour of my bill. Regretfully, despite expressions of sympathy, I was unsuccessful. Perhaps one day, someone will feel inspired to carry the torch on this issue.

CODA

DURING THE PANDEMIC, I spent more time than usual at home and enjoyed being closer to my family. I have always found the mundane pleasures of home life satisfying. I enjoy the creative process of shopping for groceries, of preparing a meal, and helping with chores. I *like* to vacuum! I enjoy walking our dog, Romeo. I was sure that adapting to home life, following my retirement, would not be difficult and I was right. This may seem anticlimactic for someone who sailed across the Atlantic twice, flew to space three times, and travelled around the world as Canada's minister of foreign affairs, but it is not. In fact, it is like that moment of peace and fulfilment after my first shuttle mission, when I had touched back down on Earth, in that I had met the challenges facing me and made my country proud.

I mentioned earlier my training in the navy as a ship's diver, describing the task of recovering a wrench at the bottom of Halifax Harbour at night, in six meters of water, ten meters from the dock, with no lights to guide me. I had only my hands to help me feel for the wrench, amidst all the garbage that had accumulated on the bottom over time. I was told not to surface until I found the tool or ran out of air.

I have often thought of that night and the sheer joy I felt when I came upon the wrench and methodically ran both my hands over it to be certain it was the wrench and not some piece of detritus. And only then did I rise to the surface. Life has many moments like that, when you're groping in the dark, facing a tough challenge. I think I can say with some pride that most of the time throughout my professional life, I found that wrench before I ran out of air.

ACKNOWLEDGEMENTS

I HAVE BEEN FORTUNATE to live a life accompanied by some special people: my parents, Jean and André Garneau; my brothers Braun, Charles, and Philippe; my first wife, Jacqueline; my children Yves, Simone, Adrien, and George; my grandchildren Elliott, Ela, and Emil; and of course my special companion of more than three decades, Pamela. My achievements, such as they are, were for them.

As I discovered during the writing of this memoir, my memory could at times play tricks on me. The differences could be subtle or, occasionally, glaring. Fortunately, my family, friends, and colleagues have helped me along the way to decipher fact from fiction.

Pamela, Simone, and Philippe read some early drafts, providing valuable suggestions and, above all, encouragement to keep writing. Marc Roy, a close friend who had been my chief of staff at Transport Canada, helped me with the chapters covering my political life. Michael Keenan, my deputy minister at Transport, reviewed those chapters with a fine-tooth comb. I am grateful to all of them.

Those who edited early drafts also encouraged me to explore the possibility of actually publishing the book commercially, rather than just keeping it to myself and my family as a vanity project. At the suggestion of my astronaut colleague Chris Hadfield, a published author himself, I approached his literary agent, Rick Broadhead, of Rick Broadhead & Associates, to test the waters. Rick took the time to read the entire manuscript (more than once), and after making a long list of suggestions, drafted a proposal that he sent to publishers.

I am most grateful to Rick for his encouragement and for getting the ball rolling, an intimidating process for a novice such as myself.

The response to the proposal was encouraging and before I knew it, we had a deal with Signal/McClelland & Stewart, an imprint within Penguin Random House Canada. Taking me on board required an act of faith on their part, and I thank them for that. Most instrumental in helping me get to the final manuscript was Doug Pepper, publisher of Signal and editor of the book. It was truly a pleasure working with him. There's no other way to put it: Doug was incredible. His editorial skills drew me out and made me dig deeper than I thought possible. I can't recount how many times he told me (I'm paraphrasing): "You told us what you did, but you didn't tell us why you did it and how you felt about it. People will want to know."

Just when I thought the editorial process was over, Doug handed the manuscript to copy editor Shaun Oakey, who meticulously reviewed every word. I cannot thank him enough for his attention to detail and for also uncovering some factual errors on my part.

I am not a natural writer. After a lifetime of drafting military memos and technical documents, I had to learn how to write a memoir that would hold the reader's attention. It's one thing to get the facts straight, it's another to write in a compelling manner. To my surprise, I enjoyed the experience.

I have not lived my life in a bubble, and there are those people whom I'd especially like to thank who mentored and guided me in my various careers. Although they did not directly assist me with writing this book, there would not be a book—at least not this one—without them. Beginning with the navy, there are those with whom I shared a profound love of the profession and who equipped me with the tools and confidence to make my own contribution, notably Norm Smyth, Jim Carruthers, Jim Dean, Cam McIntyre, and Jim Ironside. They were role models for me.

A special thanks to my fellow Canadian astronauts, whose companionship and support meant so much to me as we journeyed together, living the same matchless experiences, as well as the many talented engineers and scientists with whom I worked throughout my space career, both as an astronaut and during my time at the Canadian Space Agency. Many of them are mentioned in the book. Not mentioned were senior executives at the National Research Council who played an important role in the early days of the astronaut program: Clive Willis, Ken Pulfer, and Madeleine Hinchey.

I must also thank my extraordinary crewmates on STS-41G, STS-77, and STS-97. It was a singular honour to fly with them, and to work with NASA, an organization that inspired me every day to do my best.

Switching to politics, I want to thank the wonderful team that supported me during my bid for the leadership of the Liberal Party. I won't mention their names again, but they should know that their support meant the world to me.

I also want to thank the dedicated staffers who worked with me during my time as a minister. They gave it everything, 24/7, to ensure I was ready to face the unending challenges of political life.

At Transport Canada: Jean-Philippe Arseneau, Marc Roy, Alain Berinstain, Delphine Denis, Mélany Gauvin, Alex Jagric, Shane McCloskey, Gurveen Chadha, Heather Chiasson, Carola Haney, William Harvey-Blouin, Adel Boulazreg, Miled Hill, Philip Kuligowski Chan, Anson Duran, Amy Butcher, Maude Rousseau, Livia Belcea, Alison Murphy, Ashley Wright, Émilie Simard, Alexandre Boulé, Chris Berzins, Ally St-Jean, Laurel Lennox, Blake Oliver, Malcolm Duncan, Alexandra Scott-Larouche, Sébastien Beaupré-Huot, Helena Kojo, Richard Rock, Steve Desjardins, Marc-André Sarrazin, Justine Bellavance, Yasmine Zemni, Sidney Black, Logan Stock, Anushay Sheikh, and Alex Mendes—every one of them, a solid team player.

At Foreign Affairs: Daniel Lauzon, Ricky Landry, Elizabeth Anderson, Oz Jungic, Berit Beattie, Pierre-Yves Bourque, Samantha Nadler, Annie Lagueux, Emily Desrochers, Gabriel Vermette, Muna Tojiboeva, Olivier Duhaime, Salman Arif, Sara Amash, Sarah Manney, Shifa Tauqir, Syrine Khoury, Héléna Botelho, Nadia Hadj Mohamed, and not to forget Alexandre Boulé, Élisabeth d'Amours, William Harvey-Blouin, and Steve Desjardins, who transferred with me from Transport.

Running election campaigns is demanding. Beyond being a staffer at Transport, Alex Jagric was also my campaign manager in the 2015 and 2019 elections, and Élisabeth d'Amours, a staffer at both Transport and Foreign Affairs, was my campaign manager in the 2021 election. I thank them for getting me re-elected.

Jean Proulx deserves special mention for managing my parliamentary office by himself during the lonely years we spent in opposition and then following me to Transport to handle my legislative work.

Another key team ran my riding office, handling the requests from my constituents. I can't mention them all, but special thanks to Hervé Rivet, my first office manager in 2008, who also directed my first two election campaigns. Thanks also to François Rivet, who managed my third election campaign in 2011 and ran my office until I retired, and finally, a heartfelt thanks to Margaret Guest, my secret weapon, someone who knew everyone in the riding and understood better than anyone the issues that mattered to them.

Lastly, I must acknowledge the superb work of my parliamentary secretaries, those hardworking MPs who represent you when you're not there: Terry Beech, Chris Biddle, and Karen McCrimmon at Transport, and Rob Oliphant at Foreign Affairs: every one of them a consummate professional.

In closing, I hope I've written a book for everyone that is entertaining, thoughtful, informative, and maybe even a little surprising. Thank you!

INDEX